Bordetella pertussis

IMMUNOLOGY SERIES

Edited by NOEL ROSE

Professor and Chairman
Department of Immunology and Microbiology
Wayne State University
Detroit, Michigan

1. Mechanisms in Allergy: Reagin-Mediated Hypersensitivity
   Edited by Lawrence Goodfriend, Alec Sehon, and Robert P. Orange
2. Immunopathology: Methods and Techniques
   Edited by Theodore P. Zacharia and Sidney S. Breese, Jr.
3. Immunity and Cancer in Man: An Introduction
   Edited by Arnold E. Reif
4. *Bordetella pertussis:* Immunological and Other Biological Activities
   J. J. Munoz and R. K. Bergman

Other Volumes in Preparation

Bordetella pertussis

Immunological and Other Biological Activities

J. J. MUNOZ • R. K. BERGMAN

U.S. Department of Health, Education, and Welfare
Public Health Service
National Institutes of Health
National Institute of Allergy and Infectious Diseases
Rocky Mountain Laboratory, Hamilton, Montana

MARCEL DEKKER, INC.   NEW YORK AND BASEL

Library of Congress Cataloging in Publication Data

Munoz, John J   1918-
  Bordetella pertussis.

  (Immunology series ; 4)
  Includes bibliographies and indexes.
  1. Bacterial toxins. 2. Bordetella pertussis.
3. Bacterial antigens. I. Bergman, Robert K.,
1934-   joint author. II. Title. III. Series:
[DNLM: 1. Bordetella pertussis--Immunology.
2. Bordetella pertussis--Pathogenicity. 3. Pertussis
vaccine--Toxicity. 4. Subcellular fractions.
W1 IM53K v. 4 / QW140 M967b]
QP632.B3M86   589.9'5   76-26453
ISBN 0-8247-6507-9

COPYRIGHT © 1977 by MARCEL DEKKER, INC. ALL RIGHTS RESERVED.

Neither this book nor any part may be reproduced or transmitted in any form or by any means, electronic or mechanical, including photocopying, microfilming, and recording, or by any information storage and retrieval system, without permission in writing from the publisher.

MARCEL DEKKER, INC.
270 Madison Avenue, New York, New York 10016

Current printing (last digit):
10 9 8 7 6 5 4 3 2 1

PRINTED IN THE UNITED STATES OF AMERICA

Preface

At the suggestion of Dr. Noel R. Rose, we undertook the task of writing this book to assemble our experiences utilizing *Bordetella pertussis* vaccine or cell extracts to modify physiological, immunological, and pharmacological responses in experimental animals. Although our work is emphasized throughout, most research pertinent and significant to the topics under discussion is covered. Many areas of the *B. pertussis* field are not included.

Bordetella pertussis cells possess substances of great biological activity which undoubtedly play an important role in the pathogenesis of whooping cough. Before vaccination of children became a common practice, this disease was an important cause of death of children throughout the world. Vaccination has reduced both incidence and deaths due to whooping cough, but the vaccine has some undesirable toxic effects, which in rare cases can be fatal. For this reason, many have attempted to purify the protective antigen to develop a vaccine free of toxicity. To date, however, no one is certain about the substance or substances involved in inducing protection in children. In the process of these investigations many interesting effects of *B. pertussis* were discovered. It was found that pertussis vaccine, in addition to immunizing mice to intracerebral infection, also induces a state of hyperreactivity to histamine, serotonin, or a combination of these amines; increases susceptibility to anaphylaxis and to most forms of hemodynamic shock; increases antibody formation to protein antigens given with it; increases susceptibility to some experimental autoimmune diseases; and induces a marked lymphocytosis, an increase in plasma insulin, a hypoglycemia, and hypoproteinemia. Most, if not all, of these phenomena seem to be caused by a single substance. Various workers, including ourselves, have purified a heat labile substance which in microgram doses induces the various activities mentioned above. To avoid confusion in nomenclature we will call this substance "pertussigen" instead of referring to it as histamine sensitizing factor (HSF), leukocytosis promoting factor (LPF), or mouse protective antigen (PA). More specifically, as used in this book, the name "pertussigen" refers to a substance originating from *B. pertussis* which, although still poorly known chemically, seems to contain mainly protein and perhaps lipid. Its activities are destroyed by certain proteolytic enzymes, by heating at 80°C for 30 min, and by many protein denaturing agents. This substance is responsible for many of the biological activities of the *B. pertussis* cell. Pertussigen can be differentiated from other biologically active substances found in *B. pertussis* such as endotoxin (heat resistant), heat labile toxin (destroyed by heating at 56°C for a few minutes), and from most agglutinogens from which pertussigen has been separated. We like to think that pertussigen is "the substance" that gives *B. pertussis* its ability to produce the characteristic symptoms of whooping cough, but we realize that whooping cough results from the combined actions of all biologically active substances of the *B. pertussis* cell.

In this book we cover all the most interesting effects induced by pertussigen as well as some due to other substances. As will be seen, these phenomena are not solely bacteriological in nature, but involve many branches of the biological sciences (immunology, physiology, pharmacology, biochemistry, toxicology, pathology, endocrinology, and genetics). In complexity these problems do not differ from other biological phenomena, and we feel humble in our

PREFACE

attempts to answer many fundamental questions raised by these investigations. We present the facts and give our interpretations which hopefully will stimulate further research in this field.

Acknowledgments

Many persons have participated in the investigations made in our laboratory and we would like to acknowledge their contributions. The early work was done in collaboration with Drs. W. F. Verwey and L. F. Schuchardt who introduced one of us (JJM) to the field of *B. pertussis* and stimulated his interest to continue these studies for many years. The association with Drs. R. L. Anacker, R. F. Ross, and C. Cameron and with our former graduate students Dr. C. R. Clausen, Dr. A. Banerjea, and M. Maung provided many of the observations that have made this book possible. We would also like to recognize the help that our dedicated technicians have given us. We especially want to thank B. M. Hestekin, R. L. Cole, and J. C. Ayers who have worked with us for many years. We are grateful to Mrs. Helen Blahnik who typed the manuscript and to C. T. Taylor and R. M. Evans who made the illustrations and photographs used in the book.

    Special thanks are due to Drs. Margaret Pittman, Noel R. Rose, D. L. Lodmell, H. G. Stoenner, and H. F. Hasenclever for reviewing and suggesting many improvements to the manuscript.

J. J. Munoz
R. K. Bergman

Contents

Preface   iii

1. INTRODUCTION   1

    I. Historical Notes   1
    II. Incidence of Whooping Cough   2
    III. Symptomatology, Pathology, and Physiological Changes in Whooping Cough   6
    IV. Complications   8
    V. Experimental Observations in Animals Receiving Pertussis Vaccine   9
    References   10

2. THE *BORDETELLA PERTUSSIS* CELL AND ITS BIOLOGICAL ACTIVITIES IN ANIMALS   13

    I. General Description of *B. pertussis*   14
    II. Heat Labile Toxin   20
    III. Heat Stable Toxin (Endotoxin)   26
    IV. Agglutinogens   32
    V. Hemagglutinin   39
    VI. Protective Antigen(s)   43
    VII. Pertussigen   47
    References   65

3. SHOCK-ENHANCING EFFECTS OF PERTUSSIGEN   71

    I. Description of Phenomenon   72
    II. Factors that Affect the Enhancement of Shock by Pertussigen   73
    III. Hypotheses About Shock Enhancement by Pertussigen   87
    References   103

4. ENHANCEMENT OF ANAPHYLACTIC SENSITIVITY 109

    I. General Remarks 109
    II. Active Anaphylaxis 110
    III. Passive Anaphylaxis 114
    IV. Other Hypersensitivity Reactions 118
    V. Mechanism of Pertussigen Action in Anaphylaxis 119
    References 121

5. EFFECT OF *BORDETELLA PERTUSSIS* ON ANTIBODY PRODUCTION AND HYPERSENSITIVITY REACTIONS 123

    I. General Remarks 123
    II. Adjuvant Action of Endotoxin 126
    III. Adjuvant Effect of Whole Cells 127
    IV. Stimulation of IgE-Like Antibody 130
    V. Suppression of Antibody Formation, Delayed Hypersensitivity, and Tissue Graft Rejection 136
    VI. Mechanism of Adjuvant Action of *B. pertussis* 138
    References 139

6. EFFECT ON AUTOIMMUNE DISEASES 143

    I. General Remarks 143
    II. Hyperacute Experimental Allergic Encephalomyelitis 145
    III. Other Autoimmune Diseases 148
    IV. Mechanism of Action 148
    References 150

7. LYMPHOCYTOSIS-PROMOTING EFFECT 151

    I. General Remarks 151
    II. Recent Observations 151
    III. Mechanism of Induction of Lymphocytosis 154
    References 157

8. OTHER ACTIONS OF *BORDETELLA PERTUSSIS* 159

    I. Introductory Remarks 160
    II. Hypoglycemia 160
    III. Hypoproteinemia 162
    IV. Increased Susceptibility to Cold Stress 164
    V. Effect on Susceptibility to Infections 165
    VI. Effect on Tumors 170
    VII. Induction of Interferon 173
    VIII. Inhibition of Macrophage Response to Brain Injury 175
    IX. Hypersensitivity Reactions to *B. pertussis* 176
    References 181

CONTENTS

9. HYPOTHESES ABOUT THE MODE OF ACTION OF PERTUSSIGEN AND FUTURE WORK ON SUBSTANCES FROM *BORDETELLA PERTUSSIS* — 183

    I. Hypotheses on Actions of Pertussigen — 183
    II. Future Work — 188
    References — 189

Appendix
    MATERIALS AND METHODS USED IN OUR WORK — 191

    I. Media for Cultivation of *B. pertussis* — 193
    II. Preparation of Cell Suspensions — 194
    III. Tests for Biological Activity — 195
    IV. Methods to Measure Vascular Permeability — 198
    V. Direct Agglutination of Erythrocytes — 200
    VI. Antibody Titration by the Bis-Diazo-Benzidine (BDB) Technique — 200
    VII. Detection of Antibody-Forming Cells in Spleen and Lymph Nodes — 202
    VIII. Passive Cutaneous Anaphylaxis (PCA) in Mice — 205
    IX. Anaphylaxis in *B. pertussis*-Treated Mice — 206
    X. Preparation of Antisera — 206
    XI. Agglutination Tests — 207
    XII. Agglutinin Production in Mice — 208
    XIII. Agglutinogen Absorption Test — 208
    XIV. Immunization of Mice to Produce Ascitic Fluid Containing IgE Antibodies — 209
    XV. Enhancement of Experimental Allergic Encephalomyelitis — 210
    XVI. Schultz-Dale Reaction — 210
    XVII. Gel Diffusion Test — 211
    XVIII. Immunoelectrophoresis (IEP) — 211
    XIX. Disc Electrophoresis in Acrylamide Gel — 212
    XX. Hydroxylapatite Column Chromatography — 215
    XXI. Starch Block Electrophoresis — 215
    XXII. Zonal Density Gradient Electrophoresis — 216
    References — 216

Author Index — 219

Subject Index — 229

Bordetella pertussis

Chapter 1

INTRODUCTION

I.   HISTORICAL NOTES 1
II. INCIDENCE OF WHOOPING COUGH. 2
III. SYMPTOMATOLOGY, PATHOLOGY, AND PHYSIOLOGICAL
     CHANGES IN WHOOPING COUGH. 6
IV. COMPLICATIONS. 8
V.   EXPERIMENTAL OBSERVATIONS IN ANIMALS RECEIVING
     PERTUSSIS VACCINE. 9
     REFERENCES 10

I. HISTORICAL NOTES

Whooping cough has been an important disease of young children since it was first recorded in the medical literature in the Middle Ages (1). Thousands died and many more were left incapacitated from complications of the disease. The bacterium responsible was discovered in the sputum of children affected with whooping cough by Bordet and Gengou in 1906 (2), who also cultivated the microbe on a blood agar medium that they developed and which is still used today. They showed that suspensions of cells grown on Bordet-Gengou (B-G) agar agglutinated and fixed complement in serum of convalescent children, giving further proof of the direct relationship of the bacterium to whooping cough (3-5). For many years the bacterium was known as the Bordet-Gengou bacillus, and later as *Haemophilus pertussis*. However, in 1952 Moreno-López (6) created a special genus to encompass three serologically related Gram-negative bacilli which are associated with localized respiratory infections of man and animals. He called this genus *Bordetella* to honor Bordet. The three species of this genus are *B. pertussis, B. parapertussis,* and *B. bronchiseptica,* of which *B. pertussis* is the most important member.

The success of Pasteur in developing vaccines for various viral and bacterial diseases led many investigators to attempt development of vaccines for many infectious diseases. The discovery and cultivation of *B. pertussis* was also followed by attempts to develop an effective vaccine for whooping cough. Many unsuccessful attempts were reported before Sauer (7) and Kendrick and Eldering (8) succeeded in producing an effective vaccine. The success was largely due to the observation made by Leslie and Gardner (9) who showed that *B. pertussis* cells did not always grow in a smooth phase (phase 1) but also in a rough (phase 4) or intermediate forms (phases 2 and 3). Only smooth forms are suitable for vaccine production; when many of the difficult problems of growing, killing, detoxifying, preserving, and standardizing the cell suspensions were solved, a practical and effective vaccine was finally produced. Particularly important were the development of safety and potency standards pioneered by Pittman in this country (10).

II. INCIDENCE OF WHOOPING COUGH

The introduction of effective vaccines against whooping cough in about 1940 to 1945 resulted in a reduction of both incidence of and mortality from this disease. In Fig. 1, one can see that, even before vaccination was introduced, the death rate from pertussis declined from 1900 to 1940. This decrease in mortality must have been due to better public health practices, better nutrition, and improved medical services. After 1940, when vaccination became more general in this country, mortality rate from whooping cough showed a precipitous decline, and by 1950 mortality from this disease was almost

II. INCIDENCE OF WHOOPING COUGH

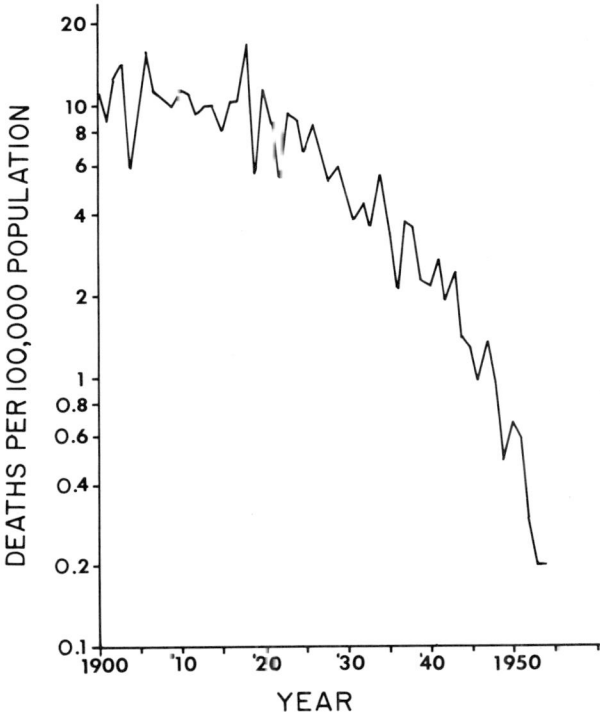

FIG. 1. Fatal cases from whooping cough per 100,000 persons in the United States from 1900 to 1952. (Data taken from Ref. 10.)

negligible (10). The reported cases of pertussis also decreased significantly after widespread vaccination was practiced (10). This is illustrated in Fig. 2 which gives the cases in the United States from 1950 to 1974 (11,12) and Fig. 3 which gives the cases in the United Kingdom from 1940 to 1959 (13). In countries where vaccination practices were started more recently, a similar dramatic decrease in the incidence of whooping cough was also noticed after introduction of pertussis vaccine. Figure 4 illustrates the morbidity rates in the Soviet Union from 1958, when vaccination was first introduced, to 1968 (14). Similar observations have been made in other countries as well. In spite of the proven effectiveness of vaccines, many countries do not have the means to purchase and the knowledge to

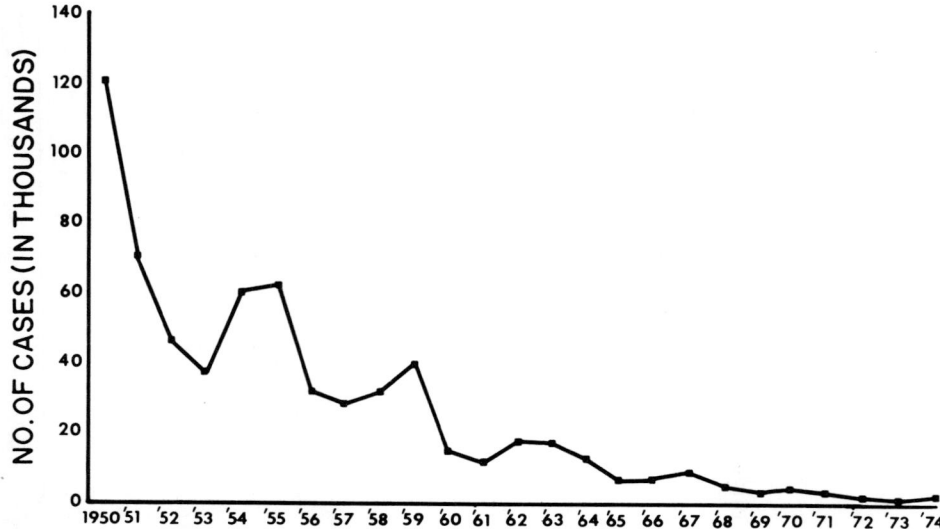

FIG. 2. Reported cases of whooping cough in the United States. (Taken from Refs. 11 and 12.)

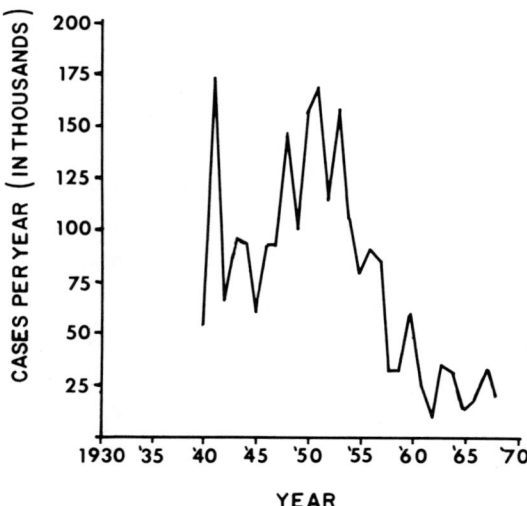

FIG. 3. Reported cases of whooping cough in the United Kingdom. Generalized vaccination introduced between 1950 and 1955. (Taken from Ref. 13.)

II. INCIDENCE OF WHOOPING COUGH

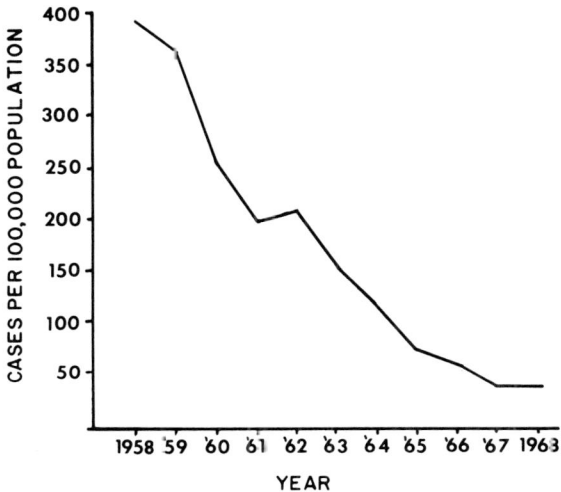

FIG. 4. Incidence of whooping cough in the Soviet Union. General vaccination was introduced in 1958. (Taken from Ref. 14.)

produce the vaccine, and the morbidity and mortality from whooping cough are still significant. The data of Table 1 taken from Raška (15) gives the average annual number of cases, deaths, and the rates

TABLE 1

Reported Cases and Deaths of Whooping Cough in the Americas (1964 to 1966)[a]

Reported accounts	North America	Central America	South America[b]
Cases (annual average)	13,166	48,909	103,511
Deaths (annual average)	79	8,366	6,070
Cases per 100,000 persons	6.2	63.1	151.3
Deaths per 100,000 persons	0.0	12.1	12.1

[a] Data taken from Ref. 15.
[b] Cases excluding Brazil and Ecuador; deaths excluding Argentina, Bolivia, Brazil, and Haiti.

of whooping cough per 100,000 persons in North, Central, and South America during the period of 1964 through 1966. Thousands of deaths still occur in Central and South America where vaccination is not generally practiced and medical care among the poor is inadequate.

III. SYMPTOMATOLOGY, PATHOLOGY, AND PHYSIOLOGICAL CHANGES IN WHOOPING COUGH

Whooping cough is a highly communicable, acute infection of the respiratory tract. After an incubation period of about 10 days, the catarrhal stage begins and lasts 1 to 2 weeks. This is followed by the paroxysmal stage with the characteristic cough that can last from 1 to 6 weeks. The convalescent or decline period ensues with a duration of 1 to 6 weeks before the child returns to normal.

In the catarrhal stage the disease cannot be clinically differentiated from other upper respiratory infections and, although coughing is usually present, no paroxysmal cough episodes are observed. Slight fever is usually found, and physical examination of the lungs rarely reveals any changes. With the onset of the second stage, 1 to 2 weeks after the first, the coughing typically develops into paroxysms which are more severe at night than during the day. These attacks are precipitated at any time but are more commonly brought on by nervous excitement, running, or inhaling moist or dusty air. When the paroxysmal stage is fully developed the symptomatology is characteristic. From 5 to 20 forceful, hacking, successive coughs, allowing little time to breathe between coughs, occur in a paroxysm. The paroxysm is often so prolonged that anoxia and cyanosis may result. These attacks vary from very mild to severe and the frequency is variable from one case to another. Vomiting is frequent during these episodes. In uncomplicated cases there is no fever during this stage. The face and upper eyelids are red and puffy. Conjunctival hemorrhages and epistaxis are frequent. Occasionally subcutaneous hemorrhages in the throat are seen. Even purpura can be observed. Examination of the nose reveals enlarged turbinates and fiery red and congested mucuous membranes. The larynx is congested

III. SYMPTOMATOLOGY, PATHOLOGY, PHYSIOLOGY

and hyperemic and the glottis on occasion shows edema. The congestion may extend to the trachea. In the lungs, large rales can be heard in the upper lobes, which as the whooping cough progresses, extend over both lungs (16).

It is interesting that in children with a history of allergic disorders a more severe and protracted disease is observed and frequently persistent bronchial asthma follows an attack of whooping cough (16).

Leukocytosis in whooping cough, first observed by Fröhlich (17), begins in the catarrhal stage. At the height of the paroxysmal stage lymphocytes comprise 60 to 80% of the total white cells. The count returns to normal within 2 to 3 months. Sauer and Hambrecht (18) found a leukopenia at the beginning and end of the disease. Lymphocytosis is a striking feature and coincides with the presence of B. pertussis in the respiratory tract (16). Leukocyte counts of as high as 192,000 per mm^3 have been reported in rare cases, and 60,000 are common with 50 to 80% lymphocytes. When the white cell count is that high, the disease may be confused with lymphatic leukemia (16). In some cases the lymphocytosis is not present.

Some workers have found the sedimentation rate to be retarded during the paroxysmal stage, but others have observed normal values (16). A significant finding by Regan and Tolstoouhov (19) was that the blood sugar levels are reduced and acidosis due to accumulation of carbon dioxide occurs.

The x-ray pictures of lungs from uncomplicated cases of whooping cough show a variable pattern of obstructive emphysema and atelectasis, depending on the amount of thick viscid mucous in the bronchial tree. B. pertussis is found among the cilia of epithelial cells lining the trachea and bronchi, and inside polymorphonuclear leukocytes free in the bronchial secretions. The bacilli were thought to damage the normal action of the cilia, resulting in an accumulation of secretions which lead to a continuous irritation and characteristic cough (20). An initial, peribronchial, lymphoid hyperplasia is observed, following or coinciding with a necrotizing inflammation of the bronchi, trachea, larynx, and nasopharynx. Often a diffuse bronchopneumonia with marked desquamation of the alveolar epithelium and lymphocytic infiltration of the peribronchial tissue and alveolar walls occurs. In areas of consolidation, alveoli filled with fibrin, leukocytes, and erythrocytes are found. The alveoli usually contain many *B. pertussis* cells (16).

IV. COMPLICATIONS

Complications are numerous in pertussis (16) and side reactions also occur after vaccination (21,22). One of the most intriguing is the neurological involvement observed in 1.5 to 14% of hospitalized cases. The most common manifestations are convulsions, coma, paralysis, blindness, and psychic disturbances. Among patients with neurological signs, mortality rates can be as high as 60 to 90% (23). These complications usually appear at the peak of the paroxysmal stage and last from a few days to several weeks. Some have reported that 1/3 of these cases recover without sequelae, 1/3 are left with varying neurological sequelae, and 1/3 remain incurable. In a followup of 35 hospitalized cases, Byers and Rizzo (24) found that 17% of hospitalized children had some permanent damage, and Schachter (25) found that even uncomplicated whooping cough apparently affected the mental capabilities of children. Neurological complications, which may cause death, have also been noticed after vaccination. According to Pittman (23), the incidence of severe reactions after vaccination in the United States is not accurately known but she estimates that not more than one fatal encephalopathy occurs in 5 to 10 million injections. In England, 1 fatal case per 1 million injections have been suspected (26). In Sweden, the incidence of all types of neurological reactions after vaccination with pertussis-diphtheria-tetanus triple vaccine is as high, according to Ström (21), as 1 per 3,600 vaccinated children.

V. EXPERIMENTAL OBSERVATIONS IN ANIMALS RECEIVING PERTUSSIS VACCINE

The introduction of pertussis vaccine for use in man was not without some practical problems, as whole cell vaccines may produce local and systemic reactions (fever, anorexia, malaise) in children. On rare occasions vaccines produce encephalitis and even death (21). These toxic reactions stimulated many pharmaceutical industries in this country and abroad to produce less toxic vaccines. The success has been limited and to this day purification and identification of the protective antigen has not been accomplished. In the course of these studies, infections with B. pertussis were induced in chimpanzees, monkeys, puppies, rabbits, rats, mice, ferrets, and chick embryos (1), and many interesting observations were made.

In Macacca mulatta and in chimpanzees, a disease similar to pertussis is produced, but in other animals the infections produced by B. pertussis do not resemble pertussis (27) and many of the deaths were most likely due to intoxication rather than infection. Resistance against intranasal (28) and intracerebral (29,30) infection of mice has been used to measure the immunizing ability of vaccines. The mouse protection test developed by Kendrick et al. (29) and by Pittman and Lieberman (30) was shown to give a good indication of the effectiveness of vaccines for children (31,32).

Studies in mice demonstrated that treatment with pertussis vaccine induced a variety of interesting changes. Among the first described was a marked increase in the susceptibility to histamine shock (33) followed shortly by the description of a number of other changes such as increased susceptibility to anaphylactic shock, serotonin, serotonin-histamine, bradykinin, peptone shock, endotoxin, anoxia, cold stress, x-rays, ether anesthesia, and perhaps many other agents (34). Furthermore, in mice pertussis vaccine induced a leukocytosis, a lymphocytosis, a decrease in blood sugar, a decrease in plasma albumin levels, and marked changes in fat and glucose metabolism (34), and increased the level of insulin in plasma (35). In addition, pertussis vaccines increased the susceptibility of mice to some infections, increased the resistance to others, and exerted a marked adjuvant effect to antigen given with it (34). Other intriguing effects of pertussis vaccine were also observed, such as an increase in IgE production with specificity to antigens given with it, and a marked increase in susceptibility to autoimmune diseases such as experimental allergic encephalomyelitis (EAE), adrenalitis, thyroiditis, and orchitis (34). *B. pertussis* cells have also been observed to increase susceptibility to certain tumors while preventing the growth of others. All these effects seem to be produced mainly through two fundamental changes induced by *B. pertussis:* (a) an increased susceptibility to shock and (b) an increased antibody production. Some of the observed changes may not be easily explained by these two activities and may be due to yet unknown mechanisms.

In this book we first describe the antigenic complexity of *B. pertussis,* the toxins produced, the agglutinogens, and the purification of some of these substances as well as their possible mode of action. We then describe the various immunological and physiological changes induced by a heat labile material which is responsible for most of the observed immunological and physiological effects of pertussis vaccine. Finally we make some speculations on the mode of action of *B. pertussis,* the possible implications of its action, and the complications that it may create. To facilitate reproduction of our work, an Appendix describing the techniques employed in our laboratory is included.

REFERENCES

1. W. L. Bradford, in *Bacterial and Mycotic Infections of Man* (R. J. Dubos and J. G. Hirsch, eds.), 4th ed., J. B. Lippincott Co., Philadelphia, 1965, pp. 742-751.
2. J. Bordet and O. Gengou, *Ann. Inst. Pasteur, 20,* 731 (1906).
3. J. Bordet and O. Gengou, *Ann. Inst. Pasteur, 21,* 720 (1907).
4. J. Bordet and O. Gengou, *Ann. Inst. Pasteur, 23,* 415 (1909).

REFERENCES

5. J. Bordet and Sleeswyk, *Ann. Inst. Pasteur, 24,* 476 (1910).
6. M. Moreno-López, *Microbiol. Española, 5,* 177 (1952).
7. L. Sauer, *J. Amer. Med. Asscc., 100,* 239 (1933).
8. P. Kendrick and G. Eldering, *Amer. J. Hyg., 29,* 133 (1939).
9. P. H. Leslie and A. D. Gardner, *J. Hyg., 31,* 423 (1931).
10. M. Pittman, *J. Wash. Acad. Sci., 46,* 234 (1956).
11. *Morbidity and Mortality, 23,* No. 53, (1972).
12. *Morbidity and Mortality,* Annual Supplement Summary, 1972, U.S. Dept. Health, Education, and Welfare, Publication No. (CDC) 74.8241.
13. F. T. Perkins, *Symp. Series Immunobiol. Standard,* Vol. 13, Karger, Basel, 1970, pp. 14-17.
14. M. S. Zakharova, *Symp. Series Immunobiol. Standard,* Vol. 13, Karger, Basel, 1970, pp. 22-23.
15. K. Raška, *Symp. Series Immunobiol. Standard,* Vol. 13, Karger, Basel, 1970, pp. 4-9.
16. J. H. Lapin, *Whooping Cough,* Charles C. Thomas, Springfield, Ill., 1943.
17. J. Fröhlich, *Jahrb. Kinderheilk, 54,* 53 (1897), Cited by Lapin (16).
18. L. Sauer and L. Hambrecht, *Amer. J. Diseases Children, 37,* 732 (1929), Cited by Lapin (16).
19. J. C. Regan and A. Tolstoouhov, *N. Y. State J. Med., 36,* 1075 (1936), Cited by Lapin (16).
20. F. Mallory and A. Hornor, *J. Med. Res., 27,* 391 (1912), Cited by Lapin (16).
21. J. Ström, *Symp. Series Immunobiol. Standard,* Vol. 13, Karger, Basel, 1970, pp. 157-160.
22. C. A. Hannik, *Symp. Series Immunobiol. Standard,* Vol. 13, Karger, Basel, 1970, pp. 161-170.
23. M. Pittman, in *Infectious Agents and Host Reactions* (S. Mudd, ed.), W. B. Saunders Co., Philadelphia, 1970, pp. 239-270.
24. R. K. Byers and N. D. Rizzo, *New Engl. J. Med., 242,* 887 (1950).
25. M. Schachter, *Praxis, 42,* 464 (1953), Cited by Pittman (23).
26. P. W. Muggleton, *Public Health, 81,* 252 (1967).
27. G. S. Wilson and A. A. Miles, *Topley and Wilson's Principles of Bacteriology, Virology and Immunity,* 6th ed., The Williams and Wilkins Co., Baltimore, 1975.
28. F. M. Burnet and C. Timmis, *Brit. J. Exptl. Pathol., 18,* 83 (1937).

29. P. L. Kendrick, G. Eldering, M. K. Dixon, and J. Misner, *Amer. J. Public Health, 37,* 803 (1947).
30. M. Pittman and J. E. Lieberman, *Amer. J. Public Health, 38,* 15, (1948).
31. Medical Research Council, Vaccination Against Whooping Cough: Relation Between Protection in Children and Results of Laboratory Tests, *Brit. Med. J., 2,* 454 (1956).
32. Medical Research Council, Vaccination Against Whooping Cough: Final Report, *Brit. Med. J., 1,* 994 (1959).
33. I. A. Parfentjev and M. A. Goodline, *J. Pharmacol. Exptl. Therap., 92,* 411 (1948).
34. J. Munoz and R. K. Bergman, *Bacteriol. Rev., 32,* 103 (1968).
35. A. Gulbenkian, L. Schobert, C. Nixon, and I. I. A. Tabachnick, *Endocrinology, 83,* 885 (1968).

Chapter 2

THE *BORDETELLA PERTUSSIS* CELL
AND ITS BIOLOGICAL ACTIVITIES IN ANIMALS

I.	GENERAL DESCRIPTION OF *B. PERTUSSIS*.	14
II.	HEAT LABILE TOXIN.	20
	A. General Characteristics	20
	B. Extraction and Purification	20
	C. Stability and Antigenicity.	21
	D. Toxicity to Various Tissues	22
	E. Role in Immunity to Whooping Cough.	24
III.	HEAT STABLE TOXIN (ENDOTOXIN).	26
	A. General Characteristics	26
	B. Extraction and Purification	26
	C. Chemical Composition.	27
	D. Antigenicity and Other Biological Activities.	27
IV.	AGGLUTINOGENS.	32
	A. General Characteristics	32
	B. Extraction and Purification	35
	C. Antigenicity and Other Biological Activities.	38
	D. Relationship of Agglutinogens to Mouse Protective Antigen and to Prophylaxis of Whooping Cough.	38
V.	HEMAGGLUTININ.	39
	A. General Characteristics	39
	B. Extraction and Purification	41
	C. Stability	42
	D. Activity on Erythrocytes.	42
	E. Role in Prophylaxis of Whooping Cough	42

VI.	PROTECTIVE ANTIGEN(S). 43
	A. General Remarks 43
	B. Attempts to Purify the Pertussis Protective Antigen . . 44
	C. Preparation of Pertussis Vaccine. 45
	D. Mouse Protective Antigen. 47
VII.	PERTUSSIGEN. 47
	A. General Remarks 47
	B. Historical Notes. 48
	C. General Characteristics 49
	D. Extraction and Purification 53
	E. Activities. 65
	REFERENCES 65

I. GENERAL DESCRIPTION OF B. PERTUSSIS

B. pertussis is a small, nonmotile, ovoid, Gram-negative bacillus (0.3 to 0.5 x 0.5 to 1 nm) which does not form spores and has a tendency to show bipolar staining. Some workers have described a capsule or sheath (1), which is not a prominent feature of most strains. Figure 1 shows that under the electron microscope well-defined capsules cannot be seen (2). In young cultures, however, a halo can usually be seen in smears treated with Gram stain (Fig. 2). Typical colonies on B-G agar have a pearly appearance, are small, grayish-white, glistening, and butyrous with smooth edges. Usually they produce a partial, diffused hemolysis on B-G agar, but this is difficult to see around young colonies. On prolonged incubation the colonies continue to grow and the hemolysis becomes more obvious. After 10 days of incubation at 37°C a more opaque center develops (Fig. 3). In liquid medium under constant agitation, B. pertussis grows as a uniform cloudy suspension, but in 2 to 3 days the culture develops a ropey, mucoid mass; most of the cells (30 to 40 x 10^9 cells/ml) remain in a smooth suspension throughout the liquid. The cells from liquid cultures are usually longer than those from B-G

FIG. 1. Electron micrograph of 28-h-old culture in B-G agar, washed several times in cold saline. Magnification = 32,785. Note the lack of clear-cut capsules on this photograph. (Reprinted from Ref. 2, p. 497, by courtesy of Williams and Wilkins Co., copyright 1959).

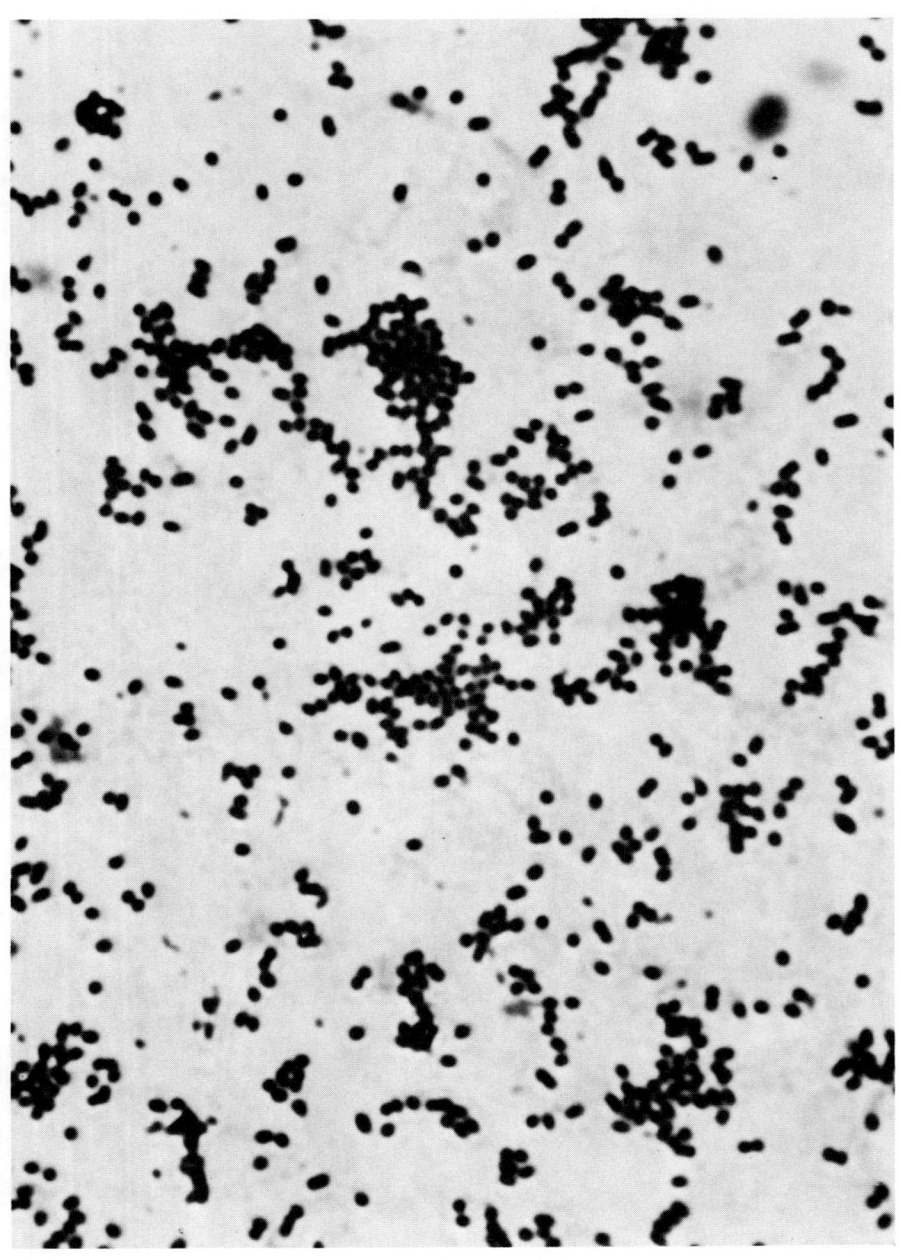

FIG. 2. Gram-stained smear made from a 24-h-old culture of *B. pertussis* grown on B-G agar.

I. GENERAL DESCRIPTION OF B. PERTUSSIS

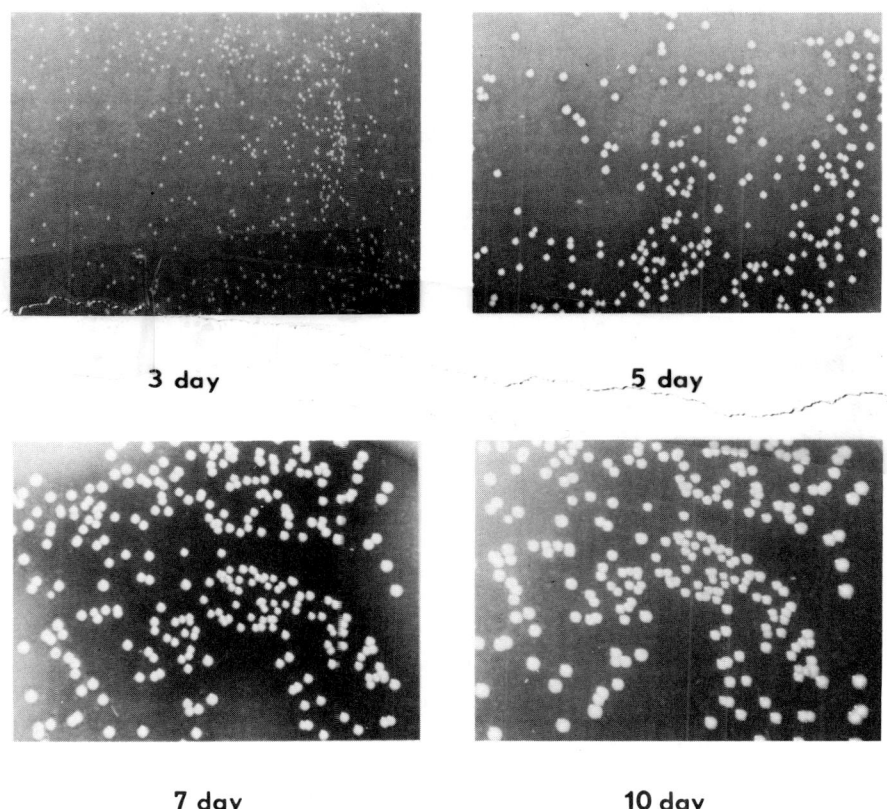

FIG. 3. Colonies of B. pertussis grown on B-G agar at 37°C for 3, 5, 7, and 10 days. Note the pearly appearance in the young cultures. A halo of diffused hemolysis is also seen in old cultures (in the black-and-white photograph this is not clearly seen). All photographs taken at a magnification of 0.76.

agar cultures. B. pertussis is aerobic, does not ferment carbohydrates, and utilizes amino acids as a source of energy. During its growth in media containing amino acids, the pH rises markedly from the release of alkaline metabolites. The growth requirements are relatively simple, but inhibitors are found in various media ingredients (such as casamino acids, yeast extract, agar) that prevent good growth. Blood, charcoal, and ion-exchange resins have been used to neutralize these inhibitors (3). The liquid media usually contain casamino acids to which various salts and soluble starch or charcoal are added. The growth conditions used in our laboratory are described in the Appendix.

Like many bacteria, *B. pertussis* has the ability to mutate from a smooth virulent form (phase 1) to a rough form (phase 4) which is nonvirulent and usually grows on plain nutrient agar. Intermediate forms (phases 2 and 3) have also been described (4). Metabolic, morphologic, and antigenic changes accompany this transformation or modulation. The smooth forms are typically coccobacilli which agglutinate in antisera to smooth cultures. They also have the ability to induce active immunity and to sensitize mice to various agents such as histamine, serotonin, and many types of shock. The rough cultures, on the other hand, usually consist of longer pleomorphic rods with many filamentous forms, which do not specifically agglutinate in the antisera to smooth forms and lack the ability to protect mice from infection or to sensitize them to histamine and various other shock-inducing substances. Some antigens found in smooth cells are absent from rough strains, but many common antigens, including the heat labile (56°C for ½ h) toxin, are retained. Figure 4, which is a gel diffusion test made with antiserum to a smooth culture in the center and extracts from smooth and rough cultures in the outside wells, shows some of the unique and common antigens found (5).

Many biologically active substances have been demonstrated in *B. pertussis*. As seen in Fig. 5, 15 different antigens have been enumerated by immunoelectrophoresis (6). At least 8 agglutinogens, a hemagglutinating substance (HA), an endotoxin (lipopolysaccharide), and a heat labile (56°C for ½ h) toxin (HLT) have been demonstrated. Furthermore, a histamine sensitizing factor (HSF), a mouse protective antigen (PA), a lymphocyte promoting activity (LPF), and a heat labile (80°C for ½ h) adjuvant activity can be demonstrated in cells from smooth cultures. These last four activities are due to one substance for which we have proposed the name "pertussigen" (7). A copper-containing pigment called azurin and many enzymes are also present. Recently some 35 different components were found by sodium

I. GENERAL DESCRIPTION OF B. PERTUSSIS 19

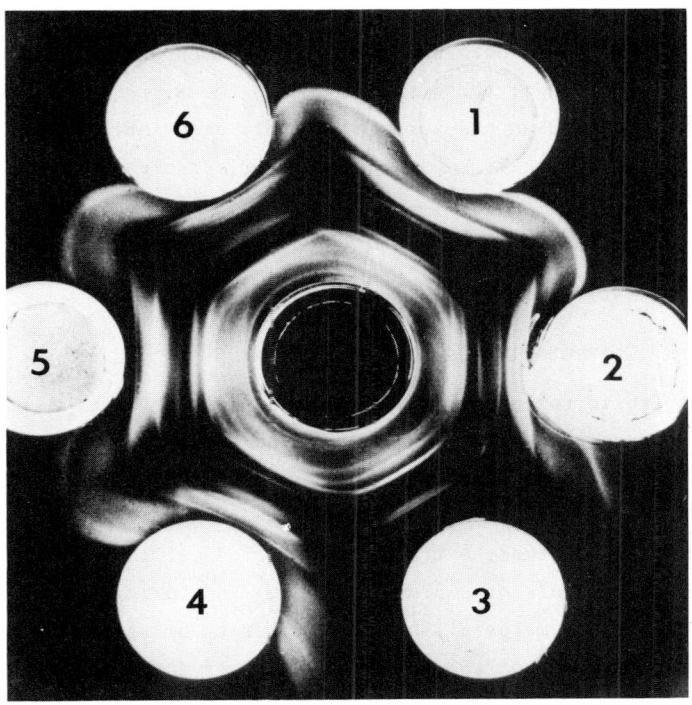

FIG. 4. Gel diffusion test with anti-B. pertussis rabbit antiserum in the center well and extracts from smooth B. pertussis cultures (wells 1, 2, 4, 5, 6) and one rough culture (well 3). (Reprinted from Ref. 5, p. 328, by courtesy of the American Society for Microbiology.)

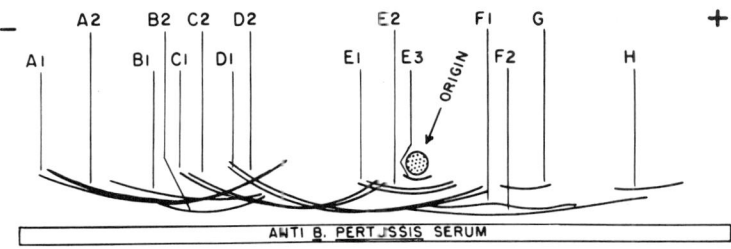

FIG. 5. Tracing of an immunoelectrophoresis test with alkaline saline extract of B. pertussis and a pool of hyperimmune rabbit sera. (Reprinted from Ref. 6, p. 59, by courtesy of the American Society for Microbiology.)

dodecyl sulfate (SDS) polyacrylamide gel electrophoresis by Parton and Wardlaw (8). No detailed studies have been done to demonstrate the exact location of these substances in the *B. pertussis* cells, but many are found on the surface probably associated with the cell walls. The heat labile substance, however, is clearly found in the protoplasm (2).

II. HEAT LABILE TOXIN

A. General Characteristics

The heat labile toxin (HLT) is a proteinaceous material in the protoplasm of young cells (2). In older, liquid-grown cells the toxin is also found in the supernatant fluid after removal of the cells. This toxin was first demonstrated by Bordet and Gengou in 1909 (9) as a dermonecrotic substance extracted from dried cells that had been ground with crystals of sodium chloride. The toxin could be collected by dissolving the material in enough water to bring the sodium chloride concentration to 7.5%. This solution of toxin given i.p. killed guinea pigs and rabbits. When given s.c. it produced a hemorrhagic edema which became a purple-black necrosis within 24 to 48 h and eventually sloughed, leaving an ulcer which later healed.

B. Extraction and Purification

The toxin can be obtained by disrupting fresh cultures with a number of cell disintegrators (10). One of the most promising methods to purify the HLT was published by Onoue et al. (11). This method as modified by Iida and Okonogi (12) briefly consists of disrupting liquid-grown cells at a concentration of 100 to 200 mg/ml in distilled water by means of a sonic disintegrator (15 kHz, 250 W) for 20 min in the cold. The sonicated cells were centrifuged at 10,000 rpm for 30 min to remove particulate matter. The supernatant fluid (250 ml) was diluted 1.5-fold with cold distilled water and was added to 50 ml of a suspension of freshly prepared calcium phosphate gel. The mixture was stirred for 15 min and then centrifuged. The pre-

II. HEAT LABILE TOXIN

cipitated gel was extracted with 90 ml of 0.1 M sodium phosphate buffer (pH 8.0) and centrifuged. To the clear extract enough saturated ammonium sulfate was added to reach 31% saturation. The precipitate was collected by centrifugation, dissolved in 10 ml of water, and dialyzed overnight at 4°C against 0.001 M sodium phosphate buffer (pH 7.4). This material was then applied to a DEAE-cellulose column (2 x 23 cm) equilibrated with 0.001 M sodium phosphate buffer (pH 7.4). Stepwise elution at pH 7.4 was performed with successive 300-ml volumes of (a) 0.005 M sodium phosphate buffer containing 0.01 M NaCl, (b) 0.02 M sodium phosphate buffer containing 0.01 M NaCl, and (c) 0.04 M sodium phosphate buffer containing 0.05 M NaCl. The toxin found chiefly in the effluent obtained with the last buffer, was precipitated by half saturation with ammonium sulfate, then collected and dissolved in a small volume of water. After dialysis against 0.005 M sodium phosphate buffer (pH 7.4), this fraction was rechromatographed in the same manner but in a smaller column. The active eluate was again precipitated with half saturation ammonium sulfate and dissolved in water. The material was then stored at -80°C.

Iida and Okonogi's preparation had an MLD for mice of 0.3 to 0.6 μg of protein, a minimal necrotizing dose of 0.005 to 0.006 μg, and reduced the spleen weight in doses of 0.07 to 0.08 μg (12). It still was not pure, however, since more than one antigen could be demonstrated by gel diffusion tests. It is a protein, since it gives maximum absorption at 277 nm, a positive test for protein and amino acids (12), and it is destroyed by digestion with trypsin but not by DNAse or RNAse (13) (Table 1).

C. Stability and Antigenicity

The toxin can easily be converted to toxoid by treatment with formalin (14-16) and in this form is a good antigen; the active toxin, on the other hand, is not (10). Antibodies to toxoid neutralize the active toxin (16). HLT is a very unstable material, losing its toxicity rapidly. At 5 to 22°C toxin solutions lose 25% of their

TABLE 1

Effect of Trypsin, DNAse, and RNAse on Toxicity of HLT
from B. pertussis[a,b]

Crude heat labile toxin (μg/mouse)	Enzyme treatment	Toxicity to mice (D/T)[c]
100	No treatment	8/9
100	Trypsin	0/9
--	Trypsin control	0/3
100	DNAse	8/9
--	DNAse control	0/3
100	RNAse	8/9
--	RNAse control	0/3

[a] Reprinted from Ref. 13, p. 272, by courtesy of the American Society for Microbiology.
[b] Each mouse received i.p. 0.5 ml test material containing 100 μg of protoplasm and either 14 μg of trypsin or 7 μg DNAse or RNAse. Controls received the untreated protoplasm or enzymes alone.
[c] Deaths/number tested.

activity within 2 days, while at 37°C 95% of the activity is destroyed in the same length of time (15). At 56°C practically all demonstrable toxicity disappears within 10 min (2). This instability has made studies with HLT difficult.

D. Toxicity to Various Tissues

The toxin does not seem to have a prominent role to play in the ability of B. pertussis to produce infection. However, the toxin produces inflammation and necrosis in the respiratory tract and these effects most likely contribute significantly to the pathogenesis of whooping cough. In chick embryos it produces lesions in the epithelial cells of the lungs (17). When given i.c., lesions are observed in the brain and meninges of guinea pigs (18). Some workers have thought that this toxin plays an important role in the encephalopathies sometimes observed in whooping cough (19).

II. HEAT LABILE TOXIN

In 1940 Wood (20) described a very interesting effect of HLT in mice. She found that mice dying from a s.c. dose of toxin showed marked injection of the vessels and small hemorrhages particularly in the abdominal musculature. In mice surviving i.p. injection of toxin and killed 1 week to 2 months later, a marked atrophy of the spleen was noticed. The spleens were small, bloodless, and frequently no more than 1/4 the normal weight. We (unpublished observations) have also seen this phenomenon and it is illustrated in Fig. 6. Wood found that histologically the spleen had almost no red pulp and few Malpighian bodies. The spaces between the remaining Malpighian bodies were filled with connective tissue. The weight of the liver was also reduced and showed a slight cellular atrophy. The abdomen was often distended with ascitic fluid and the testes were small and atrophic. Other organs looked normal. Iida and Okonogi in 1971 (12), unaware of these observations, described in more detail similar action of HLT. They observed a marked reduction of spleen weight, necrosis of

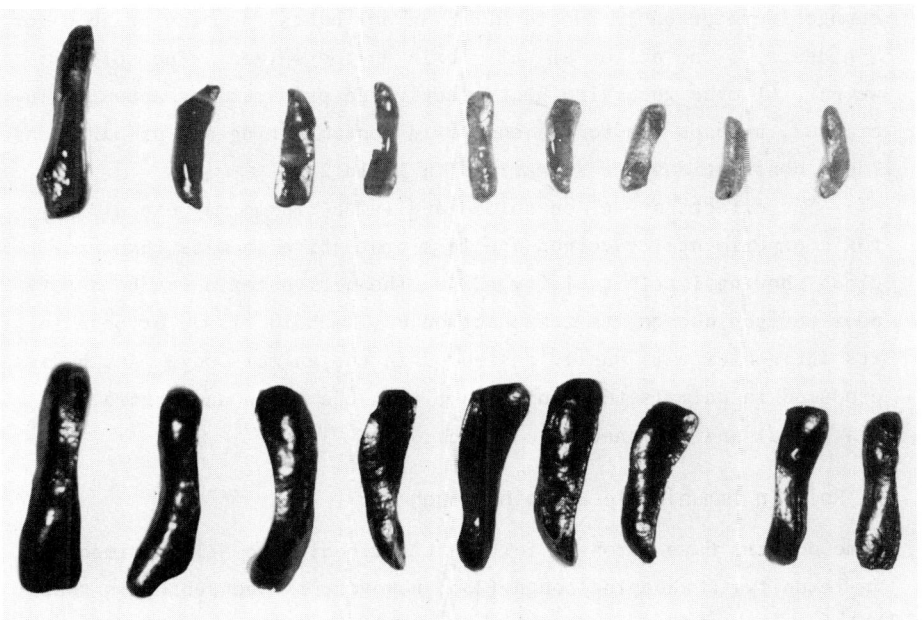

FIG. 6. Spleens of mice receiving HLT (top row) and the same amount of heat inactivated toxin (bottom row).

follicles, perifollicular hyperemia, a marked reduction of lymphocytes, fibroblastic proliferation, and fibrosis. Some changes were noticed as early as 4 h after i.v. administration of the toxin. At this time, hyperemia and edema were evident in the perifollicular areas and an indication that migration of lymphocytes into the circulation had started was also noticed. At 8 h the follicles became less distinct and at 12 h cellular degeneration was noticed, as well as some fibrosis in the perifollicular areas and a reduction of lymphocytes throughout the spleen. Two days later, marked necrotic areas were found in the red pulp, extensive fibrosis was seen, and the spleen was generally atrophic. By the 5th day the necrotic areas were reduced and fibrosis increased. On the 7th day general fibrosis of the spleen was seen and there was a virtual absence, except for an occational group, of lymphocytes. Wood found that the spleen was atrophic even 2 months after i.p. administration of toxin. Nakayama (21) has also observed toxic effect on lymph nodes. Although these studies were not done with purified toxin, most of the observed changes were probably due to HLT. In our hands, heating at 56°C for ½ h destroys the action of this toxin on the spleen. The spleen weights of mice receiving heat inactivated preparations actually increased, perhaps due to the endotoxin contaminating our preparations. These observations are summarized in Table 2.

The effects of HLT on spleen and lymph tissue may explain why toxic experimental vaccines are less protective in mice than vaccines showing little toxicity (22). The active toxin may be a very poor antigen due to its toxic action on lymphoid tissue or because its antigenicity is easily destroyed in the animal body. Antibodies produced in animals to toxoid react with the toxin and neutralize its lethal and dermonecrotic effects.

E. Role in Immunity to Whooping Cough

Some workers have strongly felt that antibodies to HLT are important in immunity to whooping cough (23); however, the consensus is that HLT is not involved in establishing active immunity to this disease

TABLE 2

Effect of HLT from B. pertussis on Body and Spleen Weights of Mice[a,b]

Unheated		Heated 56°C for ½ h		Normal spleen weight (g)
Body weight (g)	Spleen weight (g)	Body weight (g)	Spleen weight (g)	
18	0.036	20	0.156	0.109
17	0.066	19	0.159	0.132
22	0.039	22	0.216	0.129
16	0.029	21	0.170	0.153
16	0.077	24	0.135	0.101
18	0.037	22	0.154	0.113
16	0.033	23	0.144	0.140
16	0.047	17	0.164	0.178
16	0.184	19	0.124	0.114
		19	0.167	0.054
Average				
17.2	0.0609	20.5	0.1589	0.1223

[a] Data from J. J. Munoz, unpublished observations.
[b] Each mouse received i.p. 0.2 ml of 1:37.5 dilution of toxin preparation either unheated or heated at 56°C for ½ h.

since vaccine preparations free of HLT immunize mice against i.c. challenge with B. pertussis, and HLT preparations free of protective activity can also be obtained. Furthermore, the titers of antitoxin in children recovering from pertussis are low, if present. No relationship has been found between ability of B. pertussis to infect and kill mice and its ability to produce HLT (24) and some rough strains produce as much toxin as smooth strains (16).

III. HEAT STABLE TOXIN (ENDOTOXIN)

A. General Characteristics

The little work that has been done with endotoxin from *B. pertussis* indicates that this organism produces lipopolysaccharides similar in chemical composition and biological activities to endotoxin from other Gram-negative bacteria.

Among the first to show the presence of a heat stable toxin in cells of *B. pertussis* were Ehrich et al. (25) who reported a toxin that resisted heating at 100°C for 1 h. Eldering (26) independently reported that lipopolysaccharides could be obtained from *B. pertussis* by the trichloracetic acid method employed so successfully to extract endotoxins from Gram-negative bacteria by Boivin et al. (27). A somewhat more extensive work was done by MacLennan (28) who studied the lipopolysaccharides from various *Bordetella* species, especially *B. bronchiseptica*. He found endotoxin from *B. pertussis* to be pyrogenic to rabbits and toxic to mice and chick embryos. Some strains of *B. pertussis,* however, contain endotoxin with little toxicity for mice (B. Malmgren, personal communication) while others yield highly active preparations. In the limulus hemolymph assay, *B. pertussis* endotoxin is as active as endotoxins from other Gram-negative bacteria (K. C. Milner, personal communication).

B. Extraction and Purification

MacLennan (28) prepared lipopolysaccharides from *B. pertussis* by the phenol-water method of Westphal et al. (29). His preparations were not pure but had typical endotoxin properties. Ribi et al. (30) also prepared endotoxin by this phenol-water method and their preparations were freed of nucleic acids by differential centrifugation.

We have prepared endotoxin by extracting cells with trichloracetic acid as described by Kabat (31). Twenty-five grams (wet paste) of *B. pertussis* cells were suspended in 125 ml of cold distilled water and homogenized in the cold. An equal volume of 0.5 N trichloracetic acid (150 ml) was added, the mixture chilled in an

III. HEAT STABLE TOXIN (ENDOTOXIN)

ice bath and kept overnight at 2 to 5°C with constant stirring. The
suspension was then centrifuged 1 h at 2,000 x g. The supernatant
was removed and dialyzed against cold running tap water for 24 h.
To this extract 2 volumes (1,080 ml) of absolute ethyl alcohol and
20 mg of sodium acetate were added while stirring. The mixture was
kept overnight at 2 to 5°C. The precipitate was collected by centrifugation, resuspended in 60 ml water, and diayzed 24 h against 3
changes of cold distilled water, shell frozen and lyophilized. In
a lyophilized form these preparations are stable at room temperature
indefinitely. Microscopically B. pertussis endotoxin is a rod-shaped
molecule of various lengths and 50 to 100 Å in diameter. A picture
of one preparation made by Milner et al. (32) is given in Fig. 7.
B. pertussis endotoxin has been hybridized with endotoxins from
other Gram-negative bacteria (J.A. Rudbach, personal communication),
suggesting a close similarity among these endotoxins.

C. Chemical Composition

The chemical composition and biological activities of B. pertussis,
Salmonella enteritidis, and Escherichia coli endotoxins are shown
in Table 3.

Marked differences in composition are evident between the values
for B. pertussis endotoxin reported by MacLennan (28) and those of
Ribi et al. (30) which may be due to strain differences, the culture
media used, or to the different degrees of purity of the preparations.

D. Antigenicity and Other Biological Activities

The endotoxin from B. pertussis is antigenic and sera prepared against
whole cells or extracts contain easily demonstrable precipitins to
endotoxin. The endotoxins from 4 different agglutinogen type strains
of B. pertussis were found serologically identical by gel diffusion
test, but completely different from those prepared from B. bronchiseptica and B. parapertussis (6). Agar gel diffusion tests illustrating this are given in Fig. 8. B. pertussis endotoxin is a good
adjuvant to stimulate antibodies to various antigens. As little as

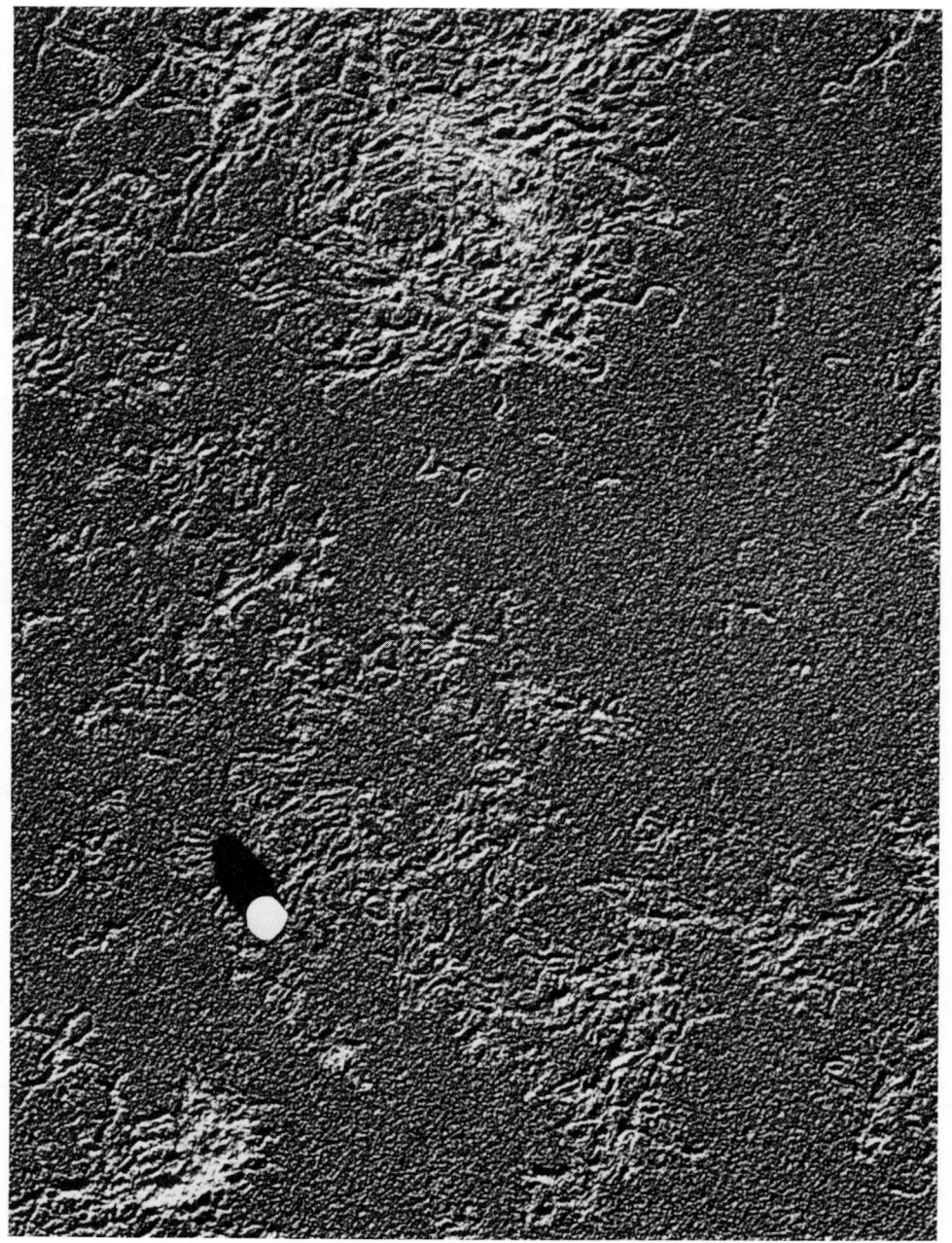

FIG. 7. Electron micrograph of endotoxin from *B. pertussis*. (Similar preparations as in Ref. 32.)

TABLE 3

Chemical and Biological Characterization of Endotoxins from Three Different Gram-negative Bacteria

Organism	N[a]	P	Hexose	Total CHO	Hexoθe amine	FAA + FAE	Heptose	DDS	KDO	Mouse, LD_{50}, i.p. (mg)	Pyrogenicity, FI_{40}, i.v. (µg)	Chick embryo, LD_{50}, i.v. (µg)
S. enteritidis[b]	2.08	1.37	54.1	61.7	3.33	9.41	1.5	13.6	2.2	ND	0.18	0.0081
E. coli 0113[b]	2.50	2.01	16.8	31.1	22.7	33.3	4.1	0	5.2	0.54	0.27	0.0049
B. pertussis[b]	3.0	2.5	9.7	15	7.3	18	9.9	0	1.25	2.1	0.64	0.0072
B. pertussis[c]	3.9	2.5	6.0	ND	20.3	ND	ND	ND	ND	>1	2.5[d]	ND

[a]Abbreviations: N = nitrogen; P = phosphorus; CHO = carbohydrate; FAA + FAE = fatty acid amide + fatty acid ester; DDS = dideoxysugar; KDO = 2-keto-3-deoxyoctulosonic acid; ND = not done or not reported.
[b]Data from Ref. 30.
[c]Data from Ref. 28.
[d]Microgram per kilogram given i.v. to rabbits was pyrogenic. No FI_{40} was reported for this fraction.

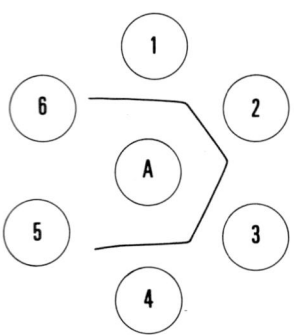

FIG. 8. Tracing of a gel diffusion test with anti-endotoxin serum (produced in rabbits by repeated injections of killed whole cells) in well A and endotoxins prepared by the trichloracetic acid method in the peripheral wells. Well 1 = B. pertussis 353/Z agglutinogen type 1; well 2 = B. pertussis 04965 agglutinogen type 1; well 3 = B. pertussis 3779 Bl_2S_4, agglutinogen type 1, 3; well 4 = B. pertussis J20, agglutinogen type 1, 2, 3; well 5 = B. parapertussis 1703, and well 6 = B. bronchiseptica 22067. (Reprinted from Ref. 6, p. 61, by courtesy of The American Society for Microbiology.)

0.5- to 5-µg doses were active in stimulating the antibody response in guinea pigs (33) and a lipid A prepared by acid hydrolysis of the lipopolysaccharide was active in doses of 15 µg. Most, if not all, of the work that has been reported on the adjuvant effect of B. pertussis has been complicated by the presence of endotoxin, although as discussed later, B. pertussis has another adjuvant that is more effective in some respects than endotoxin. All vaccines presently used for immunization of children contain endotoxin and at least some of the untoward reactions occasionally observed following vaccination are due to this substance. The role of endotoxin in immunity to whooping cough has never been adequately studied, although in the mouse experimental model, endotoxin does not induce immunity to i.c. challenge with virulent organisms. In vaccines this adjuvant most likely enhances, if nothing else, the response to the protective antigen(s). The relationship of endotoxin to the agglutinogens is not clear at present and whether it is one of the agglutinogens is still to be determined.

III. HEAT STABLE TOXIN (ENDOTOXIN)

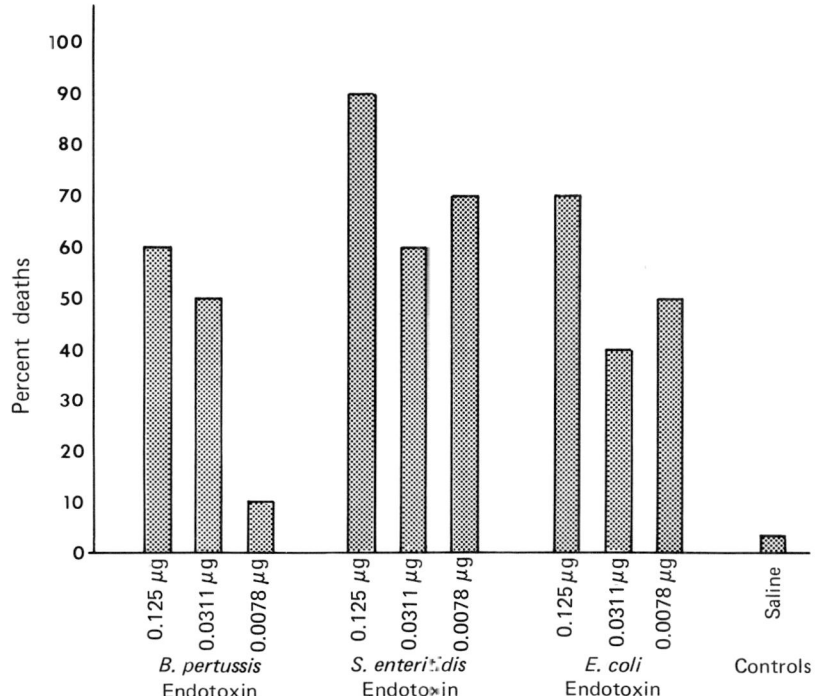

FIG. 9. Development of histamine sensitivity in mice 90 min after i.v. administration of various endotoxins. (Data from Ref. 37.)

Endotoxin has been thought to induce an increased sensitivity to histamine in mice (34,35), but the results have been erratic; large amounts were needed and a dose response was not obtained. We were not able to show increased sensitivity to histamine when endotoxin was given 3 to 4 days before challenge (36). Recently, however, we have definitely shown that endotoxin from *B. pertussis* or other Gram-negative bacteria markedly increased the susceptibility of CFW mice to histamine when challenged only 90 min after i.v. administration of minute amounts. This hypersensitivity cannot be shown when the mice are challenged 24 h or later after endotoxin administration (37). The amount required was less than 0.1 µg for *B. pertussis* endotoxin and less than 0.01 µg for *S. enteritidis* endotoxin (Fig. 9).

Endotoxin does not accelerate induction of EAE in Lewis rats. The effects of endotoxin in other phenomena described later has not been considered because most effects induced by *B. pertussis* are produced by a substance that is destroyed by heating cells or extracts at 80°C for ½ h, a treatment that does not destroy endotoxin.

IV. AGGLUTINOGENS

A. General Characteristics

The term "agglutinogen" is given to the substance or substances that react with their corresponding antibodies to cause the cells to agglutinate. They are surface antigens that can be eluted by various methods (5,39). These substances, for the most part, are known only functionally. Except for one of these agglutinogens (factor 1), little is known about their nature, although their antigenicity is useful in the classification and identification of *B. pertussis* strains.

Bordet and Sleeswyk (38) and others described the agglutination of cells by specific antisera and found that all freshly isolated cultures agglutinated in the presence of these sera. For many years the term "agglutinogen" was assumed to represent a single substance, and various workers, including our group, purified an agglutinogen which was capable of absorbing all agglutinins from an anti-*B. pertussis* serum (5), was antigenic, and failed to protect mice from i.c. infection with virulent cultures or to sensitize mice to histamine (39). Rough strains of *B. pertussis* lose their ability to agglutinate in antisera to smooth strains and concomitantly lose their activity to induce protection in mice or to sensitize them to histamine.

The antigenic complexity of the agglutinogens was not fully realized until Andersen (40) and Eldering et al. (41) showed that smooth strains have a common, heat stable O antigen and one or more thermolabile antigens. Eight different agglutinogens have been functionally described for *B. pertussis* (Table 4).

IV. AGGLUTINOGENS

TABLE 4

Analysis of Antigenic Factors of *Bordetella* Cultures
Reported by Eldering et al.[a]

Culture	Factors
B. pertussis	
5373	7, 1, 3, 6, 13
5374	7, 1, 2, 5, 6, 13
5375	7, 1, 2, 4, 13
B. parapertussis	
17-903	7, 8, 9, 10, 14
B. bronchiseptica	
5376	7, 8, 9, 12, 13
214	7, 9, 12, 13
899	7, 8, 10, 11, 12

[a]Reprinted from Ref. 42, p. 747 by courtesy of the American Society for Microbiology.

Six (agglutinogens 1-6) are species specific and two (agglutinogens 7 and 13) are shared by other species of the genus *Bordetella*. Agglutinogen 1 is common to all strains of *B. pertussis*, whereas agglutinogens 2-6 are found in various combinations as strain-specific antigens. Agglutinogens 2, 3, and 5 resist 100°C for 2½ h, but 1, 6, and 7 are destroyed by this treatment (43). Whether the various agglutinogens are separate chemical entities or are only different determinants on one molecule has not yet been resolved. The heat resistant agglutinogens likely are separate from the heat labile agglutinogens. We found that most of the factor 1 agglutinogen (heat labile) migrated to the positive pole in starch block electrophoresis (phosphate buffer, pH 6.2, μ = 0.2) while factor 3 agglutinogen (heat stable) remained entirely at the origin (44).

It is interesting that Eldering et al. (42) found that antisera specific for agglutinogen 1 could be conjugated to fluorescein isothiocyanate and that this conjugate was capable of staining *B. per-*

tussis cells, while antisera specific to agglutinogen factors 2-5 similarly conjugated were not capable of staining cells containing those factors. The agglutinating titers of the specific sera employed to make the conjugates ranged from 320 to 4,000 (Table 5). It is not clear whether the specific sera conjugated with fluorescein isothiocyanate were still capable of agglutinating the cells. These observations have been interpreted by Pittman (3) as indicating that factor 1 agglutinogen is a molecule with radicals of different specificities corresponding to the other serologic factors, but this would not explain the failure to stain with fluorescent antibody. Surely, factor 1 agglutinogen is the dominant factor in all smooth cells of B. pertussis. This agglutinogen most likely was the one studied in all the early investigations, when it was assumed that agglutinogen was a single substance (5). As Pittman suggests (3), radicals on the agglutinogen 1 could give it different specificities which would be detected as the other factors. The work of Cameron (45) indicates that factors 1-3 are found in unmodulated strains and that from these strains mutants can be obtained that may contain various combinations

TABLE 5

Comparative Results of Fluorescent Antibody Staining and Agglutination Tests with Absorbed Specific *B. pertussis* Antisera[a]

Specific serum to factor	Factors 1, 3, 7		Factors 1, 2, 5, 7		Factors 1, 2, 4, 7	
	FA[b]	agg[c]	FA	agg	FA	agg
1	4+	320	4+	160	4+	160
3	--	4,000	--	--	--	--
2 + 4	--	--	--	3,000	--	3,000
2 + 5	--	--	--	1,000	--	1,000

[a] Data taken from Ref. 42.
[b] Results of fluorescent staining.
[c] Results of agglutination with the absorbed sera expressed as reciprocal of highest dilution agglutinating the cell suspensions.

IV. AGGLUTINOGENS

of these factors (1, 2; 1, 3; or only 1). Stanbridge and Preston (46) have recently shown that these organisms can mutate in both directions, since factor 1 strains can give rise to strains having 1, 2 or 1, 3 agglutinogens or strain 1, 2 or 1, 3 can change to strains with agglutinogens 1, 2, 3. Holt (47) found that strains that were typed as containing only factor 1 when used to immunize rabbits stimulated the production of antibodies to not only factor 1 but also to factors 2 and 3.

B. Extraction and Purification

The purification of agglutinogen has been reported by various workers but no one has differentiated among the various factors or has attempted to test for the various factors in their purified agglutinogen. It can be safely assumed that all the preparations obtained contained mainly agglutinogen 1, because this is the dominant factor in all smooth cultures of *B. pertussis*. At any rate, the purified materials obtained were able to absorb all demonstrable agglutinins to the strain employed in the fractionation studies (39).

Flosdorf and Kimball (48) purified "agglutinogen" from sonically disrupted cell extracts, and later Smolens and Mudd (49) improved and simplified their technique. Intact cells were extracted with HCl at pH 1.8 at 56°C. From this extract agglutinogen was purified by a series of precipitations involving isoelectric precipitation, ammonium sulfate precipitation, and finally picric acid precipitation.

Schuchardt et al. (39) purified agglutinogens from supernatant fluids of sonically disrupted cells by adjusting the pH to 7.1 and then adding saturated ammonium sulfate until 15.25% saturation was reached. The precipitate formed at this point was removed by centrifugation and discarded. To the supernatant more ammonium sulfate was added to increase the concentration to 29.6% saturation. After 2 h at room temperature, the material was centrifuged, the precipitate washed with 30% saturated solution of ammonium sulfate, and the washed precipitate dissolved in water to 1/10 of the original volume. The agglutinogen from this solution, clarified by centrifugation, was reprecipitated by 30% saturation with ammonium sulfate.

The precipitate was collected and resuspended in 1/10 of the original volume. Table 6 gives the recovery of agglutinogen and the relative degree of purification obtained. It is clear that, with the strain employed, the agglutinogen could be completely released from the cells by sonic treatment and that from these soluble preparations agglutinogen could be precipitated by 30% saturation with ammonium sulfate. The units of agglutinogen per milligram of nitrogen increased from 4,571 to 2,509,804 after the first precipitation with ammonium sulfate. The reprecipitation did not increase the specific activity but removed traces of contaminants.

Various other attempts to purify agglutinogens have been reported by others (50) but the most detailed and complete study was made by Onoue et al. (51). The method employed by them follows:

Cells grown in liquid medium were collected by centrifugation, washed with 0.15 M NaCl, and the cell paste resuspended in saline with 0.01% merthiolate and kept at 4°C. The pH of the suspension was adjusted to 3.7 with 1 N HCl and centrifuged at 4,000 rpm for 15 min. The packed cells were ground with glass powder in a ball mill and extraction was carried out for 17 to 20 min. The extracts were pooled and centrifuged at 18,000 rpm for 20 min. To the supernatant fluid 1 N NaOH was slowly added to bring the pH to 7. A precipitate that formed was collected by centrifugation and washed 3

TABLE 6

Agglutinogen Activity and Nitrogen Content of Purified Agglutinogen[a]

Preparation	Agglutinogen (units/ml)	Nitrogen (mg/ml)	Agglutinogen (units/ml N)
Cell concentrate	12,800	2.800	4,571
S x S (supernatant)	12,000	0.819	14,652
Agglutinogen (10x)	128,000	0.051	2,509,804
Agglutinogen (10x) (reprecipitated)	64,000	0.031	2,064,516

[a] Data taken from Ref. 39.

IV. AGGLUTINOGENS

times with water at pH 7. The washings were combined with the extract and designated crude extract. An equal volume of saturated ammonium sulfate was added to the crude extract and allowed to stand overnight at 2 to 4°C. The precipitate was collected by centrifugation, washed once with 1/2 saturated ammonium sulfate, and dissolved to 1/10 of the original volume of crude extract. It was now dialyzed for 48 h against 3 changes of 2 liters of distilled water each and any insoluble precipitate was removed by centrifugation at 10,000 rpm for 15 min. The dialyzed solution of the precipitate was cooled to 0°C and methanol added slowly with stirring to a final concentration of 13%. The solution was cooled to -5°C, allowed to stand for 30 min, and centrifuged at 10,000 rpm for 7 min. Methanol was added slowly to the supernatant to a final concentration of 25%, allowed to stand at -5°C for 30 min, and centrifuged as before. The precipitate was dissolved in distilled water and dialyzed in cold distilled water overnight. This material was cleared by centrifugation and lyophilized. This agglutinogen was then passed through a DEAE column made in 0.3 M potassium phosphate buffer, pH 7.7. After the column was made it was washed with 0.005 M sodium phosphate, pH 7.7. About 200 mg of the methanol precipitated agglutinogen in 5 to 10 ml of 0.005 M sodium phosphate buffer, pH 7.7, were placed on the column. Buffer was passed to remove nonabsorbing materials and then 0.05 M sodium phosphate buffer at pH 7.1 was added. The first protein peak obtained by this elution contained highly active agglutinogen. Rechromatography of this peak resulted in a single sharp peak when eluted with 0.04 M sodium phosphate buffer, pH 7.1. The recovery in this peak had 98% of the protein rechromatographed. This material moved slowly in the ultracentrifuge and had a molecular weight of 10,000. A faster moving contaminant was seen and a trace contaminant was detected by agar diffusion test. The uv absorption spectrum showed an optimum absorption at 275 nm. The nitrogen content was 14.76% and no phosphorus or nucleic acids were found. The authors concluded that this agglutinogen was a simple protein. The serotype of this agglutinogen was not given.

C. Antigenicity and Other Biological Activities

Agglutinogen as purified by Onoue et al. (51) was antigenic and produced specific skin reactions of induration and erythema in sensitized rabbits. The reaction reached its maximum at 20 to 24 h. This agglutinogen was not toxic to mice in 5-mg doses given i.p.

Agglutinogen preparations have been used as skin test reagents to detect susceptibility to whooping cough in children and to evaluate the efficacy of vaccines. Flosdorf et al. (52) found that 99% of children from 6 to 14 months of age who had no history of whooping cough had negative skin tests, whereas 75% of those with history of having had the disease had positive skin tests to agglutinogen. The positive skin test correlated with serum agglutinins. The skin test, however, was later found unreliable and has not been used in public health practice.

D. Relationship of Agglutinogens to Mouse Protective Antigen and to Prophylaxis of Whooping Cough

Since the development of the mouse protection test for evaluating pertussis vaccine (53-55), many efforts to identify the protective antigen have been made. According to the British Medical Research Council (56), the protective potency of vaccines for children correlates well with the mouse protective activity. However, it cannot be stated categorically that the same substance is responsible for inducing immunity in mice and children. Until the substances are tested in pure form in both children and mice this question will remain unanswered. A relationship has been found between agglutinins in children and their resistance to whooping cough (56) but this is to be expected when complex vaccines are used since the children would develop antibodies against the various antigens found in the vaccine. Purified agglutinogen (factor 1) has not been effective in protecting mice from i.c. challenge (39) and strains of different agglutinogen types do not differ significantly in immunizing or in killing mice immunized with factor 1 strains (57,58). By passive transfer test, Eldering et al. (59) did not find any association of serotypes with mouse protection. Infection in mice, in spite of some similarities (3), is not comparable to the clinical disease in children. The ability of the mouse to destroy B. pertussis given by routes other than i.c. suggests a resistance not shared by nonimmune children. When B. pertussis is given i.c., however, it multiplies profusely and kills mice when the density of bacteria reaches a critical point, usually within 10 days (3). The blood-brain barrier, which excludes serum antibodies and phagocytic cells, must prevent the destruction of B. pertussis. Vaccines may do more than just induce antibody production. As shown later, they may have a dual role in mouse protection by enhancing antibody formation and by increasing vascular permeability (60).

Preston (61,62) has emphasized the role of agglutinogens in inducing protection in children. This may well be the case, but this point has not been conclusively settled.

The agglutinogen composition of strains isolated in England has changed (63); in the late 1940s and mid-1950s the dominant serotype was 1, 2, 4, while in the late 1950s and early 1960s the dominant serotype was 1, 3. In 1963 to 1964 only 15% of 132 strains had agglutinogen 2. Changes in serotype have also been noticed in Canada (64), Australia (65), and the United States (66). According to Preston (62), this is significant in the prophylaxis of pertussis, since vaccines which do not contain agglutinogen 3 are, in his hands, inferior to those containing 1, 2. Stanbridge and Preston (67) have developed a marmoset protective test that they claim is specific for agglutinogen factors and could prove of value in evaluating the potency of vaccines.

V. HEMAGGLUTININ

A. General Characteristics

Hemagglutinin (HA), a substance derived from B. pertussis with the ability to directly agglutinate erythrocytes from chickens and other animals, was first described by Keogh et al. (68,69). They found this substance in cell-free supernatant fluids from broth cultures of *B. pertussis*. HA was mainly found in cells of very young cultures grown on solid or in liquid media; as liquid-grown cultures grew older, most of the HA was found in the cell-free supernatant fluid. The relationship, as found by Masry (70), between HA production and growth curves of *B. pertussis* is shown in Fig. 10. HA concentration in the supernatants was maximal after 8 days of incubation, at which time the cells themselves had very little activity. It is also noteworthy that by the 12th day the HA activity had almost disappeared from both cells and fluids.

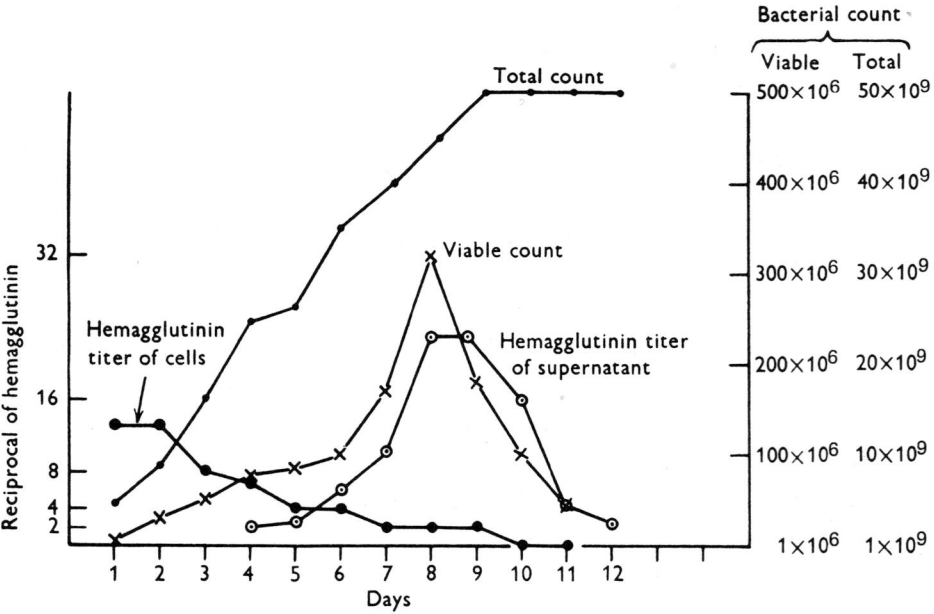

FIG. 10. Relationship between HA and growth of *B. pertussis* in liquid medium. (Reprinted from Ref. 70, p. 203 "Production, Extraction and Purification of the Haemagglutinin of *Haemophilus pertussis*" by courtesy of Cambridge University Press.)

B. Extraction and Purification

HA was extracted by Masry (70) by two different methods: In one method the cells grown for 24 h in B-G agar were harvested, suspended in 2 M NaCl to a density of 4×10^{-10} organisms/ml, and incubated at 37°C for 48 h. The suspension was centrifuged at 3,000 rpm at 0°C. The supernatant fluid contained the HA, of which the average titer of these extracts was 1/128.

The second method that yielded purer preparations was as follows: Organisms (8×10^{10} cells/ml) were suspended in 1 M sodium acetate for 48 h at 37°C. The cells were separated by centrifugation and the extract, which agglutinated red blood cells in dilutions of 1/128 to 1/256 was cooled to -3°C and the pH adjusted to be slightly acid. Methanol, chilled to -18°C, was then added slowly (about 2 ml/min) with constant stirring (the temperature of the mixture was maintained at -5°C), until the final concentration of 40% (v/v) was reached. The mixture was allowed to remain at -10°C overnight and then it was centrifuged at the same temperature for 1 h at 5,000 rpm. The supernatant was discarded and the precipitate resuspended in distilled water at 2°C and lyophilized. The dried material was readily soluble in 0.5 M phosphate buffer, pH 6.8. This precipitation with methanol purified the material quite considerably since an extract with a titer of 1/128 of the original material had 0.7 mg N/ml while the precipitated material diluted to the same titer contained only 0.02 to 0.03 mg N/ml.

Masry (70) found that these methanol precipitated preparations did not contain agglutinogen (probably factor 1), did not protect mice from infection, but were antigenic in rabbits. Intradermal injections of the purest preparations produced necrotic lesions similar to those induced by the heat labile toxin. When given i.v., they killed mice. The toxicity of extracts, however, was not directly proportional to the HA titer and adsorption of the HA with a 10% erythrocyte suspension at 37°C completely removed HA but not toxin. Thus, the toxicity observed was not due to HA.

C. Stability

Partially purified HA was completely destroyed after a few minutes at 60°C, after 72 h at 37°C, and after 4 days at room temperature. At 4°C, the titers fell 25% in 2 weeks and after 35 days HA was not detected. The rate of deterioration was not influenced by concentrations of 0.3% formalin, 0.5% phenol, or 1/10,000 thimerosal. Fifty percent glycerine, however, retarded the rate of deterioration. HA was almost completely absorbed during filtration through Seitz filters. Filtration through Gradocol filters reduced the activity only slightly (70).

Fisher (71) found that heating HA preparations at 56°C for 16 min destroyed the hemagglutinating capacity, but not its ability to adsorb to red cells or its antigenicity in rabbits.

D. Activity on Erythrocytes

HA acts on human O, A, B, and AB erythrocytes as well as on chicken, mouse, guinea pig, rabbit, sheep, and horse cells, while ox erythrocytes are not agglutinated unless pretreated with trypsin. Treatment of HA with diastase destroyed its activity against human and chicken erythrocytes, but papain or trypsin treatment destroyed only the agglutinating activity to human, but not chicken erythrocytes (72). Thiele (73) found it to be nondialyzable.

HA binds to different receptors on erythrocytes than those reacting with influenza virus (74).

E. Role in Prophylaxis of Whooping Cough

Some of the original observations indicated that HA was responsible for active protection against experimental infection in mice (7) and by inference against whooping cough, because antibodies to HA were found in sera of convalescent children (71). Others (70,73,75) seem to have excluded the role of HA in mouse protection against i.c. challenge. Thiele (73), for example, was able to remove HA by repeated absorptions with erythrocytes while the mouse protective antigen (PA) remained in the supernatant. Various other workers have

VI. PROTECTIVE ANTIGEN(S)   43

failed to induce active protection with semipurified HA, and anti-HA has also failed to induce passive protection in mice (70).

Some of the properties of HA differ from those of the protective antigen. For example, HA is very sensitive to heat (56°C for 16 min) while the protective activity is resistant. The steps employed to purify HA, however, could also concentrate protective activity and HSF, i.e., 1 M NaCl facilitates extraction of these activities and pH 4.6 precipitates both of them. No doubt, all the HA preparations so far made are heavily contaminated with other substances from the cell and for this reason little significance can be attached to studies made with impure preparations. Recently Sato and Arai (76) purified the lymphocytosis promoting factor (LPF) from extracts of agar on which *B. pertussis* had grown. Details of this procedure are given in Section VII. These extracts agglutinated erythrocytes from various animals, promoted lymphocytosis, protected mice from infection, and sensitized mice to histamine. They inferred that the four activities are part of a single substance. In our hands, HSF preparations do not have a high hemagglutinating activity, but our purification procedure is different and alteration of this activity may take place. LPF attaches readily to erythrocytes (76,77), but Morse and Morse (78,79) think that the filamentous material seen by electron microscopy is the HA and the LPF is associated with a particulate substance.

Sato and Arai's (76) preparations of LPF seem to be relatively pure, however, and further work is necessary to know the exact relationship of HA to LPF, protective activity, and HSF. This is discussed further in Section VII.

VI. PROTECTIVE ANTIGEN(S)

A. General Remarks

The discovery of the whooping cough bacillus (80) started a quest for an effective vaccine against pertussis. Many attempts were made with variable results (81). Potent vaccines were obtained when it was realized that recently isolated smooth cultures were effective and that these cells often lost their smooth properties by subculturing on artificial media. The pioneer work of Sauer (82) and that of Kendrick and Eldering (83) established the basic factors for preparation of pertussis vaccines. The development of a mouse protection and toxicity test gave two important guidelines to evaluate the potency and safety of vaccines (3). The experience throughout the world attests to the effectiveness of the presently used vaccines (see Chapter 1) and various well-controlled clinical trials have proven the value of whole cell vaccines, as well as some semipurified preparations of the *B. pertussis* cell (56,84). Both morbidity and mortality rates have been decreased by vaccination (3).

The exact nature of the substance or substances involved in inducing active immunization of children by these vaccines is not known. The mouse protection test correlates well with the efficacy of whole cells or crude preparations of soluble vaccine in protecting children from whooping cough (56). This correlation, however, does not prove that the antigens involved are the same, and whether immunity in man to pertussis is induced by the same substance(s) as those that protect mice from the artificial intracerebral infection. Preston (61) believes that agglutinogens are all important in inducing active immunization in children. All active vaccines induce production of agglutinins, and Preston's arguments are difficult to dispute with the evidence presently at hand. Only clinical trials in children immunized with single purified preparations of the various substances or with various combinations of these substances will answer this question. Furthermore, the feasibility of removing all toxins from a vaccine and still retain its potency can only be answered by this type of investigation. The mouse test cannot answer these questions.

B. Attempts to Purify the Pertussis Protective Antigen

Most efforts to obtain a protective antigen free from other biological activities of the *B. pertussis* cell have been unsuccessful.

VI. PROTECTIVE ANTIGEN(S)

Pillemer et al. (85) prepared what they thought was a highly purified protective antigen, which was effective in children, but it contained histamine sensitizing activity for mice and induced production of agglutinins. We purified the so-called HSF and found it to contain PA (86). Niwa (87), Pieroni et al. (88), Sato et al. (89), Lehrer et al. (90), and Parker and Morse (91) have obtained highly purified preparations of either HSF or LPF and most concluded that their preparations had high mouse protective activity and that all of these activities were probably found in the same substance. We have recently improved on our purification of pertussigen (7) and have again found that the purest preparations induce histamine sensitization, induce lymphocytosis, protect mice from i.c. infection, and have many other activities which are discussed later.

C. Preparation of Pertussis Vaccine

As already indicated, the presently used vaccines are effective in the control of pertussis. The United States (92) and other nations and the World Health Organization (93) have specified requirements for potency and toxicity of vaccines which are followed by manufacturers of pertussis vaccine. The relative safety of vaccines attest to the wise selection of the control measures imposed on all manufacturers. Although most "trade secrets" are not fully disclosed, potent vaccines are prepared generally as follows: A culture in smooth form (phase 1) is selected to contain all the major agglutinogen factors (1, 2, and 3). This culture is propagated in a medium as free of antigenic materials as possible. A casamino acid base (see Appendix) is usually employed, although different manufacturers modify or have developed different media which under their conditions support adequate growth. Some use large fermentation tanks with effective aeration devices, others use continuous cultivation (94), while others employ liquid medium in 1- to 2-liter Blake bottles with constant shaking. Solid charcoal-agar medium is also used by some. In all cases, blood or serum are not used because of their high antigenicity.

Some have devised synthetic media which in their hands have produced adequate growth and cells with adequate potency (95).

The stock cultures of B. pertussis should be kept in a lyophilized form to avoid mutations which could render the culture inadequate to produce potent vaccines. The lyophilized cultures are first grown on a solid medium such as B-G agar or preferably on an agar medium free of blood. From the growth on these media small volumes (200 ml) of liquid medium are inoculated and after 2 to 3 days of incubation larger volumes of media are heavily inoculated. The cultivation of cells should be as short as possible and generally never longer than 3 days at 37°C in each transfer. After the cells have grown to the desired concentration (in our hands this is usually 30 to 40 x 10^9 cells/ml), the cells are collected by centrifugation or by precipitation by lowering the pH to 4 and allowing the precipitate to settle (96). The supernatant medium is discarded and the cells resuspended to a desired concentration in physiological saline containing 0.01 to 0.02% thimerosal. The cells are heated at 56°C for 30 min to destroy the heat labile substance and then tested for their mouse protective potency, mouse toxicity, safety, and sterility before diluting to the accepted concentration for human use. The potency is expressed as units compared to the national or international standard pertussis vaccine. The killing and detoxification procedures vary, but thimerosal and heat are satisfactory at least under laboratory conditions. Some workers employ 0.1 to 0.2% formalin. However, this is somewhat deleterious to the mouse protective activity of vaccines (97). Other substances have been used, but less satisfactory results are obtained (3).

Pertussis vaccine is usually used with alum, aluminum hydroxide, or aluminum phosphate as an adjuvant, and frequently it is mixed with tetanus and diphtheria toxoids and also with poliomyelitis vaccine.

The aluminum hydroxide or phosphate gels improve the mouse protective potency of pertussis vaccine, although it is not clear whether this adjuvant improves the potency of the vaccine in man (96).

Pertussis vaccine, however, serves as an adjuvant to the antigenicity of the toxoids in mixed vaccines (98,99). In commercial production of pertussis vaccines, many problems are encountered, such as variability of growth, potency, and toxicity of the cells, as well as variability of the cultures and stability of the mouse protective activity. These problems are not within the scope of this book and are not discussed further. Those interested in these problems can find much information in the International Symposium of Pertussis (published in *Symp. Series Immunobiol. Standard.*, Vol. 13, Karger, Basel, 1970).

D. Mouse Protective Antigen

The mouse protective substance (PA) in *B. pertussis* cells is closely associated with the histamine sensitizing and the lymphocytosis promoting activities. The purest preparations of any of these substances have been found to possess all 3 activities. The most important mouse protective substance is pertussigen, which is described in the following section.

VII. PERTUSSIGEN

A. General Remarks

The name "pertussigen" has been proposed (7) for the substance from *B. pertussis* cells that induces physiologic and pharmacologic changes in mice leading to an increased susceptibility of these animals to many types of shock such as anaphylactic, histamine, serotonin, cold, x-ray, endotoxin, anoxia, etc. Furthermore, this substance induces lymphocytosis and hypoglycemia; it acts as an immunological adjuvant when given with an antigen increasing production of antibodies of various classes including the IgE class of immunoglobulins in the rat and in the mouse. Pertussigen also accelerates induction of EAE in Lewis rats receiving encephalitogenic antigen and it protects mice from intracerebral challenge with virulent *B. pertussis*. All of these activities are markedly reduced by heating pertussigen at 80°C for ½ h.

B. Historical Notes

The ability of pertussis vaccine to protect mice against intracerebral challenge was studied by Pittman (53) and Kendrick et al. (54) who developed the protection test for evaluating potency of pertussis vaccine. In 1948 Parfentjev and Goodline (100) observed that mice receiving B. pertussis vaccine became about 100-fold more susceptible to histamine than did normal mice (Table 7). These observations were soon confirmed by many workers (5) and later it was found that mice became more susceptible also to serotonin, bradykinin, endotoxin, peptone shock, active and passive anaphylaxis, anoxia, cold stress, x-rays, etc. (5). In Table 8 the magnitude of the increased susceptibility to some of these agents is given.

TABLE 7

Effect of B. pertussis on Sensitivity of Mice to Histamine[a]

Histamine diphosphate (mg/mouse)	Histamine base equivalent	Normal mice D/T[b]	%D	B. pertussis-sensitized mice D/T	%D
0.5	0.18	--	--	2/6	33
1	0.36	--	--	5/8	63
2	0.72	--	--	12/15	80
5	1.81	--	--	8/8	100
20	7.25	0/10	0	--	--
50	18.12	5/6	83	--	--
100	36.23	5/5	100	--	--

[a] Data from Ref. 100.
[b] Deaths/number tested. B. pertussis-treated mice received i.p. 75 x 10^9 killed cells 5 days before i.p. challenge with the histamine dose indicated.

TABLE 8

Increased Sensitivity to Various Agents After Administration of Pertussigen in Mice

Substance	Increased sensitivity	References
Histamine	50 to 300-fold	100-105
Serotonin	20 to 50-fold	105-107
Passive anaphylaxis	3.5 to 7.5-fold	108, 109
Endotoxins	3 to 100-fold	104, 110

C. General Characteristics

Pertussigen is produced only by smooth cells of B. pertussis and is associated with the cell wall of young cells (Table 9). In older cultures the activity is found in the media as well (76). It is not known whether it is an integral part of the cell wall structure or associated with it and released as the cell dies. Pertussigen has not been identified with any of the agglutinogens, the HLT, or endotoxin. Sato and Arai (89,111) think that the HA, HSF, PA, and the LPF are identical. We have made preparations of pertussigen that do not agglutinate erythrocytes to a significant degree and the work reported in the previous section seems to indicate significant differences between pertussigen and HA, and recently Morse and Morse (79) found that purified LPF did not have hemagglutinating activity.

The biological activities of pertussigen are almost completely destroyed by heating the preparation at 80°C for ½ h and some inactivation occurs even at 60°C. Formaldehyde (0.2%) at 37°C also reduces the histamine sensitizing and mouse protective activities markedly within 7 days, although at 2 to 5°C the effect is not striking. The protective activity is more resistant to formalin than is

TABLE 9

HSF and Mouse Protective Activity of Whole Cells, Cell Walls, and Protoplasm of B. pertussis Cells[a]

Preparation	SD_{50}[b] (μg/mouse)	PD_{50}[c] (μg/mouse)
Whole cells	14.8	8.6
Cell walls	<10	6.9
Protoplasm	33	25.6

[a] Data from Ref. 2.
[b] SD_{50} = dose of preparation sensitizing 50% of the mice to the lethal effect of 0.5 mg of histamine given i.p. 4 days later.
[c] PD_{50} = dose protecting 50% of the mice challenged i.c. with virulent B. pertussis cells 14 days after immunization.

the histamine sensitizing activity (97) and complete inactivation of HSF with retention of protective activity has been reported (89). Its ability to enhance EAE is also inactivated by heat and by formalin (112). In the presence of 0.1 to 0.2% thimerosal, pertussigen retains its activities. In a lyophilized form pertussigen is stable even at room temperature.

Pertussigen is negatively charged at pH 7.2 or above, and at pH 6 it does not move in starch block electrophoresis (5). It precipitates at pH 4.5; extensive dialysis in water also precipitates it (87).

Alcohol in the cold or acetone at room temperature does not inactivate pertussigen, but shaking in chloroform destroys its activity.

Pertussigen is most soluble at high pH (8 to 9) and high salt concentration (0.4 to 1 M NaCl); it precipitates at 35% saturation with ammonium sulfate and is insoluble in distilled water or in very low salt concentration. It adsorbs to red cells and lymphocytes (77) and most likely to many particles because large losses in activity are observed by passing it through Sephadex, starch, cellulose (filter paper), agarose, agar, and Sepharose (J. J. Munoz, unpublished

VII. PERTUSSIGEN

observations). It is found in different molecular aggregates depending on the salt concentration used (113). According to Sato and Arai (76), it is made up of molecules 4 x 20 nm as determined by electron microscopy [but Morse and Morse (79) do not agree with this observation]. In its soluble form in a high salt concentration or in 4 M urea and 1 M NaCl, it has a molecular weight of 86,000 to 108,000 (76,90) and a sedimentation coefficient of 5 to 5.5 S. As will be seen later, there are inconsistencies in the chemical composition reported by various workers. These discrepancies may be due to contaminated preparations or to differences in the chemical structure of pertussigen from different strains of B. pertussis. Except for Morse and Morse's preparations (79) all have been found to contain protein, lipid, and some carbohydrate. With proteolytic enzymes (pronase, trypsin, papain) we have not completely destroyed the activity of pertussigen (Table 10). Sato et al. (111) obtained similar results; however, at higher concentrations of pronase, HSF and LPF activities have been markedly reduced (114). Some, though not striking, inactivation of HSF activity was produced by trypsin, diastase, and lipase. Others have also reported inactivation by proteolytic enzymes (5). RNAse, DNAse, amylase, cellulase, and lysozyme do not inactivate pertussigen (5). It is only slowly inactivated by $NaIO_4$ and thus carbohydrate probably is not involved in its activity (115-117).

Purified pertussigen is tolerated by mice in i.v. or i.p. doses of over 10 µg, although some workers have found it more toxic (11, 118). Our preparations were made from acetone-treated cells, and possibly this treatment reduces toxicity without affecting other biological activities. Also, pertussigen from different strains of B. pertussis may differ slightly in their toxicity. Given in the skin or footpads, 0.5 µg induces an edematous (swelling) reaction that disappears by 24 h. In mice a toxicity that is not acute but delayed has been noticed (118). The general properties of pertussigen are summarized in Table 11.

TABLE 10

Effect of Some Enzymes on HSF Activity[a,b]

Enzyme	Concentration of enzyme in substrate (mg/ml)	SD_{50} of BPE[c] (μg)			% activity after treatment
		Enzyme	Heated enzyme	No enzyme	
Lipase	0.5	19.8	11.7	--	59
Trypsin	1	32.2	12.9	11.6	40
Papain	1	5.94	8.92	--	150
Pronase	1	12.6	4.97	4	39
Pronase	1.98	5.8	0.814	0.699	12
		15.7	0.855	0.699	5
Ficin	1	7.94	12.4	--	156
Diastase	5	29.7	15.3	25	52

[a]Data from J. J. Munoz, unpublished observations.

[b]All enzymes were tested in the recommended buffers and their activity controlled by appropriate substrates. A control with enzyme destroyed by boiling for 3 to 5 min was always included. The percent of the activity left after enzyme treatment was calculated by dividing SD_{50} value obtained with heated enzyme by the value obtained with active (unheated) enzyme and multiplying this by 100.

[c]BPE = *B. pertussis* extract made with 1 M NaCl and 0.05 M sodium pyrophosphate at pH 8.5.

TABLE 11

General Properties Reported for Pertussigen

1. Probably protein
2. Polydispersed molecules
3. Insoluble in water
4. More soluble in high salt and alkaline pH
5. Precipitates from saline solution at pH 4.5
6. Inactivated at high (above 9) and low (below 3) pHs
7. Inactivated by formalin
8. Stable in frozen or dried state
9. Merthiolate does not affect its activities
10. Inactivated by heating at 80°C for 30 min
11. Inactivated by various solvents such as chloroform
12. Some proteolytic enzymes (pronase) at high concentrations markedly reduce its activity
13. Molecular weight 86,000 to 108,000
14. Many activities in mice; it increases susceptibility to shock, it is an adjuvant, affects physiological functions, and produces many pharmacological alterations.

D. Extraction and Purification

Soluble preparations of pertussigen can be obtained from culture supernatant fluids of old cultures; by disintegrating cells by sonic treatment, sodium deoxycholate or lysozyme; by extracting them with mixtures of thiourea-urea and formamide, or 4 M urea and 1 M NaCl; or by pretreating the cells with acetone and then extracting them with alkaline saline and by other methods (5,36). These soluble preparations have served as starting materials for further purification of pertussigen.

Some properties of pertussigen complicate its purification. As purification proceeds, pertussigen becomes less soluble; it adsorbs to cells and many particles (5,77); it is inactivated by many substances (phenol, chloroform, etc.), and at pH >9 or pH <3. A large proportion of pertussigen in the cell does not become soluble and is lost in cell residue. Furthermore, biological assays used to measure activity are not quantitatively dependable. In spite of these limitations, reasonably pure preparations have been obtained. Niwa (87) obtained a high degree of purification by precipitating culture supernatant fluids with zinc acetate at pH 6 to 6.2. The precipitate was then extracted with 20% $Na_2HPO_4 \cdot 12H_2O$ and dialyzed against water at pH 6.2. The precipitate that developed during dialysis contained the active material. It was dissolved in 0.1 M phosphate buffer at pH 8 and dialyzed against water for 24 h at 4°C. The precipitated material had an SD_{50} of 0.46 µg per mouse.

We obtained purified preparations by subjecting alkaline saline extracts of B. pertussis cells to starch block electrophoresis at pH 6.2 in 0.024 ionic strength phosphate buffer. Under these conditions, pertussigen migrated only slightly toward the positive electrode and was separated from most of the demonstrable antigens in crude extracts (86). Pieroni et al. (88) extracted pertussigen from acetone-extracted cells by suspending them in water and breaking them with glass beads in a Waring blender. After centrifugation to separate the cell debris, the active material was precipitated by 35% saturation with ammonium sulfate. About 96.6% of the nitrogen of the original extract was removed by 2 precipitations with ammonium sulfate. The SD_{50} of this product was from 3.7 to 7.1 µg of nitrogen.

Morse et al. (79,91,119) obtained purified LPF from culture supernatant fluids of B. pertussis cells grown in liquid medium. After the cultures had been incubated a few days, thimerosal was added to a final concentration of 0.02% and the cells were removed by centrifugation. Culture supernatant fluids were precipitated by 90% saturation with ammonium sulfate. The precipitated material was washed with water and with 0.15 M NaCl and finally dissolved in 0.1 M Tris-0.5 M NaCl, pH 10. The soluble material was layered on a discontinuous CsCl gradient (6 ml of each of the following densities: 1.5, 1.3, 1.25 and 1.2) and centrifuged in an SW.25 rotor at 50,000

VII. PERTUSSIGEN

x g for 3.5 h. The bulk of LPF was found in the top 6 ml load volume which was then dialyzed against Tris-NaCl buffer. This fraction was finally sieved through Sephadex G-150 in a 2.6 x 26 cm siliconized column. Three fractions were obtained (I, II and III) with the bulk of the LPF activity in fraction II while the hemagglutinating activity was found in fraction I. The substance in fraction II had a molecular weight of around 87,000. The purified LPF when subjected to immunoelectrophoresis gave only one band against various antisera to B. pertussis, and when used to immunize rabbits, a single antibody was obtained. This purified LPF appears to contain 4-5 polypeptide subunits and at least 17 amino acids. It contained 14.5% nigrogen, 47.9% carbon, and 6.5% hydrogen. As little as 0.02 µg of LPF protein given i.v. to mice induced a significant leukocytosis 3 days later and in some strains of mice as little as 0.01 µg of LPF protein given i.v. sensitized them to histamine. LPF also interfered with the hyperglycemic response to epinephrine in mice. Under the electron microscope LPF appears as ring shaped particles 75 to 80 Å in diameter. Each was composed of four to five round suburits measuring 28 Å in diameter.

Sato et al. (76,89,111) purified the LPF from supernatant fluids obtained from either liquid or from agar in which B. pertussis had grown. By 60% ammonium sulphate precipitation, starch block electrophoresis, and sucrose density gradient centrifugation they obtained a highly purified fraction which contained not only LPF, but also HSF, hemagglutinating activity, and mouse protective activity (89,111). Chemically, this substance contained 7.5% nitrogen, 46.8% protein, 24.5% sugar, 4.5% amino sugar, 17.5% lipid, 1.8% phosphorus, and no DNA or RNA. The substance appeared to be uniform by ultracentrifugation and gave one band against antiserum to this substance. By electron microscopy it appeared as uniform filamentous molecules of about 2 x 40 nm. The sedimentation coefficient was 5.54 S and a molecular weight of 108,000.

Lehrer et al. (90,120) have extracted HSF from cells by means of 4 M urea-1 M NaCl buffered with 0.1 M sodium phosphate at pH 6.0-

6.4. This extract was then passed through Bio Gel A and then on Sephadex G-100. Further purification was obtained by ion exchange chromatography through a Sulphopropyl (SP) Sephadex G-50 column, from which the HSF was eluted by a continuous pH gradient (pH 4.0-4.8). The purified material in submicrogram doses given i.v. produced a marked leukocytosis, had adjuvant activity, induced lung edema, reduced the size of the thymus and lymph nodes, and increased the size of the spleen. Mice receiving 6 μg showed some loss in weight.

We (7; unpublished observations) have obtained pertussigen from acetone extracted and dried (ADC) cells by suspending them overnight in 1 M NaCl containing 0.05 M sodium pyrophosphate at pH 7.4. The

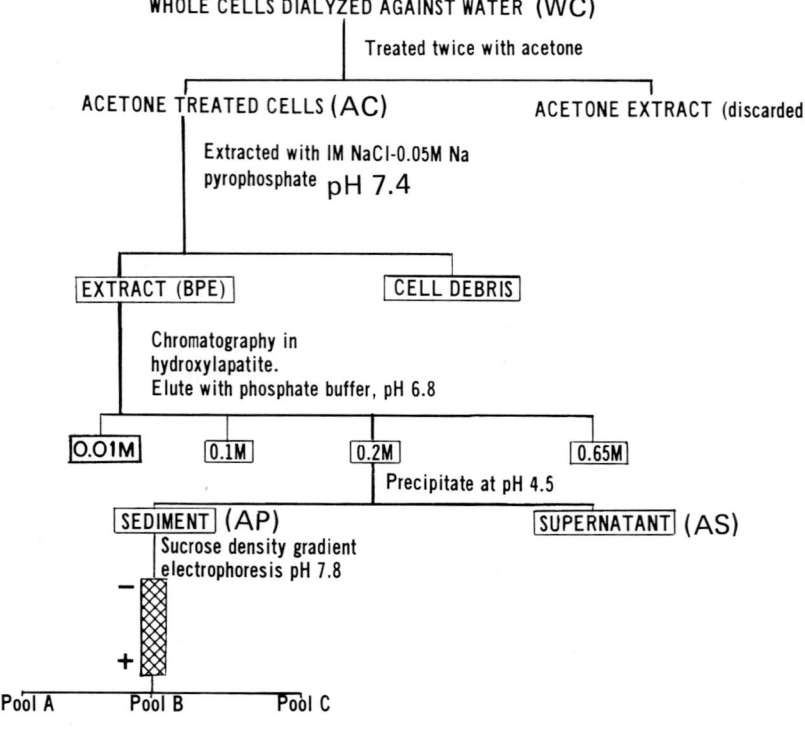

FIG. 11. Outline of fractionation steps in the purification of pertussigen. (Taken from J. J. Munoz, unpublished observations.)

VII. PERTUSSIGEN

suspension was centrifuged at 27,000 x g for 1 h and the supernatant fluid dialyzed against water. The pertussigen could then be adsorbed on an hydroxylapatite column in 0.01 M phosphate buffer, pH 6.8. At this phosphate concentration all of the HSF activity was retained by the column. The column was then treated with 0.1 M buffer and then with 0.2 M buffer which eluted most of the activity. The active fraction could be further purified by precipitation at pH 4.5 (AP)(see Fig. 11). This material has many of the activities ascribed to pertussis vaccine as can be seen in Table 12. This acid-precipitated material was the most active fraction in all the tests. The purification obtained is clearly shown in Fig. 12 which is a disc electrophoresis of the various fractions and Fig. 13, which is an immunoelectrophoresis test developed with antiserum to the starting material. Chemical analysis of this still impure fraction has shown it to contain 7.01% nitrogen, 2.5% phosphorus, and 12.6% fatty

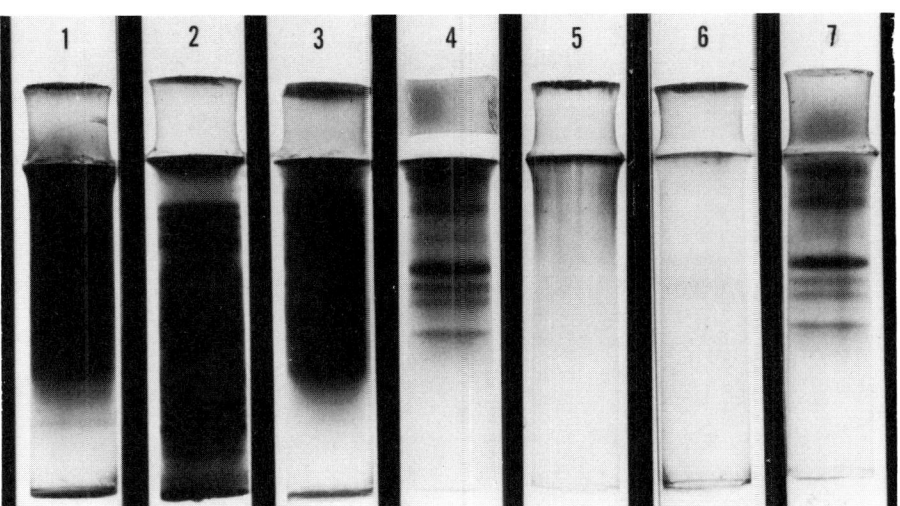

Fig. 12. Disc acrylamide gel electrophoresis of various soluble fractions obtained in the purification of pertussigen. 1 = BPE, 2 = 0.01 M, 3 = 0.1 M, 4 = 0.2 M, 5 = 0.65 M, 6 = AP, and 7 = supernatant fluid after acid precipitation. Fractions with activity = 1, 4, 6; fractions with little or no activity = 2, 3, 5, 7. (Taken from J. J. Munoz, unpublished observations.)

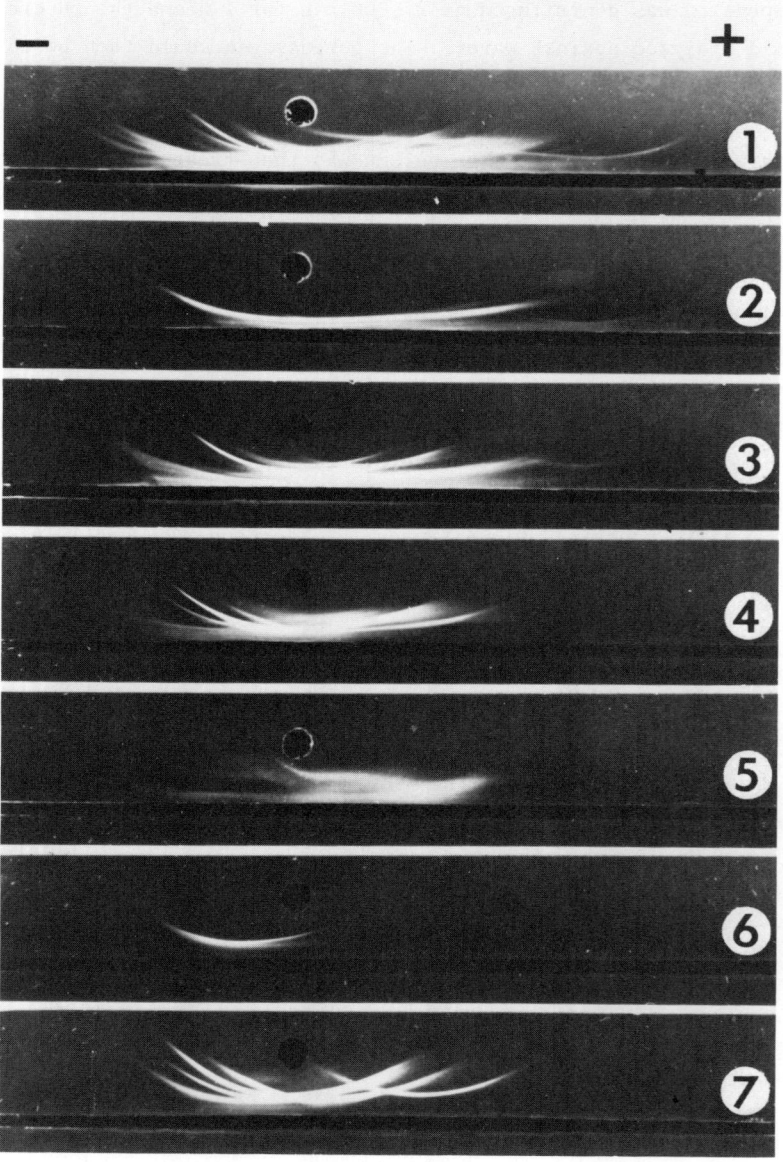

FIG. 13. Gel immunoelectrophoresis of various soluble fractions obtained in the purification of pertussigen. 1 = BPE, 2 = 0.01 M, 3 = 0.1 M, 4 = 0.2 M, 5 = 0.65 M, 6 = AP, and 7 = supernatant after acid precipitation. The antiserum placed in the trough was a pool of hyperimmune rabbit sera. (Taken from J. J. Munoz, unpublished observations.)

VII. PERTUSSIGEN

TABLE 12

Some Activities Known to be Induced by AP

I. Increases sensitivity to shock
a. Histamine
b. Serotonin
c. Active anaphylaxis
d. Passive anaphylaxis
II. Acts as an adjuvant to stimulate antibody production
a. Increases response to protein antigens
b. Stimulates 72-h PCA antibodies (IgE class)
c. Stimulates 2-h PCA antibodies (IgG_1 class)
III. Increases susceptibility to EAE in Lewis rats
IV. Induces immunity to *B. pertussis* infection
V. Induces various physiological effects
a. Leukocytosis with a marked lymphocytosis
b. Hypoglycemia
c. Hypoproteinemia

acids. The nitrogen content of this material is as low as any that has been reported. A comparison of the chemical composition of various preparations reported by the various workers is shown in Table 13. It is clear from these data that wide discrepancies are found in the reported values.

The acid-precipitated material can be further fractionated by zonal gradient electrophoresis at pH 7.8 in a 10 to 40% sucrose gradient in 0.03 M phosphate buffer. The electrophoresis is performed in an LKB electrophoresis apparatus as described by Svensson (121). The current is applied for 16 h at 600 volts and about 16 to 18 ma. Typical results obtained are illustrated in Fig. 14 which shows the optical density (280 nm) of the eluted fractions. It is clear that 3 distinct substances can be separated from acid precipitate. Fraction A is a fast-moving substance containing RNA. It does not precipitate with *B. pertussis* hyperimmune serum and does not have any HSF or protective activity. B is a middle, broad

TABLE 13

Comparison of Chemical Composition of Various Preparations of Pertussigen[a]

Reports	N	P	C	Protein	CHO	Amino sugar	Lipid	DNA	RNA	Mol wt in daltons
Sato and Arai (76)	7.5	1.8	--	46.8	24.5	4.5	17.5	0	0	108,000
Morse and Bray (119)	14.5	0.6	51.2	74 - 80[b]	3.8	1.2	1.0	<0.3	<0.3	--
Niwa (87)	5.7	2.8	--	--	2.1	<0.4	--	0	0	--
Lehrer et al. (90)	--	--	--	30.5	2.1[c]	--	57.0	<2	10.4	86,000
Munoz et al.[d]	7.01	2.5	--	--	2.9	0.21	12.6	--	--	--
Morse and Morse (79)	14.5	--	47.9	100	<1	--	<0.5	--	--	87,000

[a]Values for chemical analysis given as percent.
[b]Amino acid.
[c]Reducing sugars.
[d]Unpublished work. Analysis of acid precipitated pertussigen (AP) described above.

VII. PERTUSSIGEN

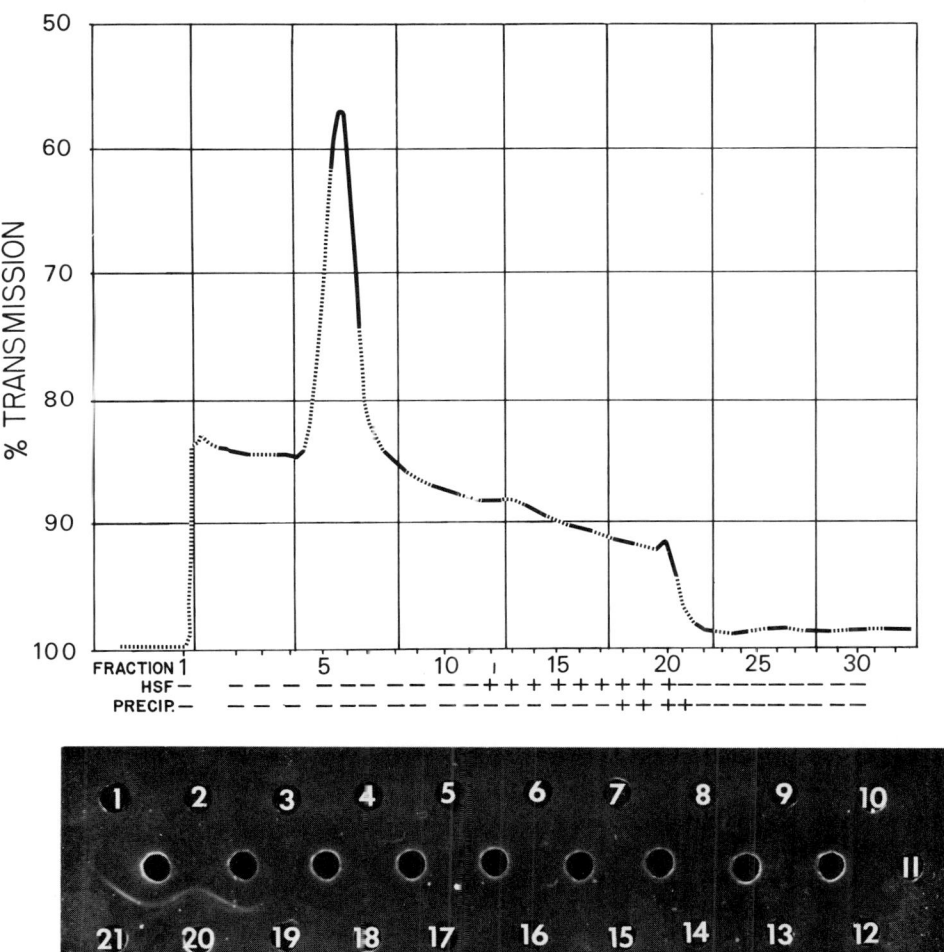

FIG. 14. Percent transmission at 280 nm of sucrose density gradient electrophoresis fractions (top) and gel diffusion test on all fractions (bottom) of AP. Pertussigen was mainly found in fractions 12 to 20 (pool B), while the precipitin bands were found only in fractions 18 to 21 (pool C). The main absorbing peak at the left was due to inactive nucleic acids (pool A). (Taken from J. J. Munoz, unpublished observations.)

fraction that contains no precipitating activity against *B. pertussis* hyperimmune serum but possesses most of the HSF activity, as well as the protective and EAE-inducing activities. Fraction C contains electrophoretically neutral substances, precipitates with anti-*B. pertussis* serum, and has some HSF and other biological activities, probably due to contamination with B. Fraction B is the most highly purified preparation of pertussigen we have obtained. Table 14 gives some of the activities of this preparation.

It should be emphasized that some important differences exist between all the preparations considered to be highly purified by the various workers. As seen in Table 13, the nitrogen, phosphorous, protein, carbohydrate, lipid, and RNA contents vary considerably. This may mean that all the materials have been contaminated and that the true chemical composition of pertussigen is still to be resolved. Our preparation has only weak hemagglutinating activity, and we have failed to demonstrate precipitating activity with potent anti-*B. pertussis* serum, but its antigenicity is presently under study.

Other workers have also purified this substance but have not achieved better purification. Iwasa et al. (118) purified lymphocytosis promoting factor by taking young culture supernatant fluids of *B. pertussis* cultures to which 0.01% thimerosal had been added and kept a few days at room temperature. The cells were removed by centrifugation, the LPF precipitated by saturation of the supernatant with ammonium sulfate and allowed to stand overnight. The precipitate was suspended in saline (0.15 M NaCl) to 1/100th of its original volume and dialyzed in saline. The insoluble material was collected by centrifugation and extracted 3 times at room temperature with 0.02 M Tris buffer, pH 9, containing 0.2 M NaCl. The combined extracts were then precipitated by adding ammonium sulfate to 70% saturation. The precipitate was dialyzed against saline and the resulting material was extracted 3 times with 0.02 M Tris buffer, pH 9, containing 0.4 M NaCl. The combined extracts were centrifuged at 100,000 x g for 60 min and the supernatant concentrated by vacuum dialysis to a concentration of 0.03% protein. This was passed through

VII. PERTUSSIGEN

TABLE 14

Some Activities of Purified Pertussigen (Fraction B)

Activity measured	μg/mouse[a]
Histamine sensitization	0.260[b]
Leukocytosis	0.368[c]
Stimulation of IgE	0.680[d]
EAE	6.800[e]
Mouse protection	1.800[f]

[a] Concentrations were determined by the Folin-Ciocalteau method. The concentrations were obtained from a standard curve made with bovine serum albumin.

[b] Amount given i.v. 3 days before challenge which induced sensitization of 50% of the mice (SD_{50}).

[c] Amount given i.v. that 3 days later elevated the leukocyte count by at least 2-fold.

[d] Amount that when given i.p. mixed with 125 μg of hen's egg albumin (HEA) induced at least a 10-fold increase in 72 h PCA antibody (IgE) titer as compared to mice receiving HEA only.

[e] Amount that when mixed with 200 mg (wet weight) of guinea pig spinal cord emulsion and given i.p. to Lewis rats induced paralytic EAE in about 50% of the rats. Rats receiving cord only seldom develop any symptoms.

[f] Amount that when given i.p. to mice protected 50% of the animals (PD_{50}) from intracerebral infection with *B. pertussis*.

a G-100 Sephadex column equilibrated in 0.02 M Tris buffer in 0.4 M NaCl, pH 9. About 50% of the LPF was found in the second peak. The pooled, active fractions were rechromatographed under the same conditions and a single protein peak was obtained containing nearly 100% of the activity placed on the column. This fraction produced a single line against anti-LPF serum and none against anti-whole cell rabbit sera with high agglutinin titer. Anti-LPF neutralized its activity. It had a sedimentation coefficient of about 5 S. This material was destroyed by trypsin and pronase at 39°C for 60 min and was also completely inactivated by 0.2% formalin which converted it to a highly antigenic toxoid.

In summary we can say that the purification work on pertussigen is not yet definitive.

The most impressive papers on the purification of pertussigen are those of Sato and Arai (76) and Morse and Morse (79), who claimed purity to the point of obtaining homogeneity by immunological and physical criteria. Since there are marked differences in the chemical analyses of the products obtained by these two groups the actual composition of pertussigen is still unsettled. Sato et al. found almost 25% carbohydrate, 17.5% lipid, and only 46.8% protein, while Morse and Morse (79) found it to contain only protein. As mentioned above we have not been able to correlate a precipitin band with pertussigen, although both Sato et al. and Morse and Morse have. The reason for our failure may well be due to concentration of pertussigen. However, we have also failed to show strong neutralization of histamine sensitizing activity by the various antisera. Only two antisera thus far tested have shown neutralization of HSF activity in soluble preparation of pergussigen when minimal amounts of pertussigen were used and when the serum concentrations were high. Another discrepancy is the hemagglutinating capacity. In our hands, and in Morse's, hemagglutinating activity is not significant. We have tested one of Sato's semipurified preparations and found it, as he did, highly active in hemagglutinating chicken and rabbit erythrocytes. The hemagglutinating activity of our cultures falls very rapidly with time of incubation. It is possible that these differences are due to the strain of *B. pertussis* employed, or to the medium employed. If this is the case, it is possible that pertussigen may be a family of compounds differing in composition and activities from one strain to another. It is clear that even if Sato's and Morse's preparations are pure as claimed, more work is needed to establish the true identity of pertussigen in various strains. Our purification studies reported here are not final and we need to improve mainly the yields of pertussigen so as to allow us to collect enough material to do meaningful chemical studies.

E. Activities

Most of the interesting immunological activities of B. pertussis cells are due to pertussigen (see Table 12). These activities affect not only the immunological responses of the animal but also many physiological, endocrinological, and pharmacological reactions. The interrelationship of these effects are a fascinating aspect of this problem and one that has occupied much of our time. The remainder of this book is devoted to describing each of the best known activities of pertussigen.

At the risk of oversimplification, we assume that pertussigen is responsible for most of these effects, although it is recognized that much of the work has been done with very crude materials containing various biologically active substances from the B. pertussis cell.

REFERENCES

1. G. M. Lawson, Amer. J. Diseases Children, 46, 1454 (1933).
2. J. Munoz, E. Ribi, and C. L. Larson, J. Immunol., 83, 496 (1959).
3. M. Pittman, in Infectious Agents and Host Reactions (S. Mudd, ed.), W. B. Saunders Co., Philadelphia, 1970, pp. 239-270.
4. P. H. Leslie and A. D. Gardner, J. Hyg., 31, 423 (1931).
5. J. Munoz, Bacteriol. Rev., 27, 325 (1963).
6. R. Ross, J. Munoz, and C. Cameron, J. Bacteriol., 99, 57 (1969).
7. J. Munoz, Fed. Proc., 35, 813 (1976).
8. R. Parton and A. C. Wardlaw, J. Med. Microbiol., 8, 47 (1975).
9. J. Bordet and O. Gengou, Ann. Inst. Pasteur, 23, 415 (1909).
10. J. Munoz, in Microbial Toxins (S. Kadis, T. Montie, and S. Ajl, eds.), Vol. IIA, Academic Press, Inc., New York, 1971, pp. 271-300.
11. K. Onoue, M. Kitagawa, and Y. Yamamura, J. Bacteriol., 86, 648 (1963).
12. T. Iida and T. Okonogi, J. Med. Microbiol., 4, 51 (1971).
13. A. Banerjea and J. Munoz, J. Bacteriol., 84, 269 (1962).
14. P. Teissier, J. Reilly, E. Rivalier, and H. Cambasse, J. Physiol. Pathol., 27, 549 (1929).

15. D. G. Evans and H. B. Maitland, *J. Pathol. Bacteriol.*, *45*, 715 (1937).
16. M. E. Roberts and A. G. Ospeck, *J. Infect. Diseases*, *71*, 264 (1942).
17. M. Gallavan and E. W. Goodpasture, *Amer. J. Pathol.*, *13*, 927 (1937).
18. P. Fonteyne and J. Dagnelie, *Compt. Rend. Soc. Biol.*, *110*, 978 (1932).
19. R. K. Byers and F. C. Moll, *Pediatrics*, *1*, 437 (1948).
20. M. L. Wood, *J. Immunol.*, *39*, 25 (1940).
21. H. Nakayama, *J. Keio Med. Soc.*, *36*, 1238 (1959), Cited in *Jap. Med.*, Vol. 1, Abst. No. 3176 (1961).
22. M. Pittman, *J. Infect. Diseases*, *89*, 300 (1951).
23. L. P. Strean, D. La Pointe, and E. Dechene, *Can. Med. Assoc. J.*, *41*, 326 (1941).
24. A. F. B. Standfast, *J. Gen. Microbiol.*, *5*, 250 (1951).
25. W. E. Ehrich, A. Bondi, S. Mudd, and E. W. Flosdorf, *Amer. J. Med. Sci.*, *204*, 503 (1942).
26. G. Eldering, *Amer. J. Hyg.*, *36*, 294 (1942).
27. A. Boivin, I. Mesrobeanu, and L. Mesrobeanu, *Comp. Rend. Soc. Biol.*, *114*, 307 (1933).
28. A. P. MacLennan, *Biochem. J.*, *74*, 398 (1960).
29. O. Westphal, O. Luderitz, and F. Bister, *Z. Naturforsch.*, *7*, 148 (1952).
30. E. Ribi, R. L. Anacker, R. Brown, W. T. Haskins, B. Malmgren, K. C. Milner, and J. A. Rudbach, *J. Bacteriol.*, *92*, 1493 (1966).
31. A. Kabat, *Kabat and Mayer's Experimental Immunochemistry*, 2nd ed., Charles C. Thomas, Springfield, Ill., 1961.
32. K. C. Milner, R. L. Anacker, K. Fukushi, W. T. Haskins, M. Landy, B. Malmgren, and E. Ribi, *Bacteriol. Rev.*, *27*, 352 (1963).
33. J. R. Farthing and L. B. Holt, *J. Hyg.*, *60*, 411 (1962).
34. R. E. Pieroni, E. J. Broderick, and L. Levine, *J. Bacteriol.*, *91*, 2169 (1966).
35. S. Malkiel and B. J. Hargis, *J. Allergy*, *35*, 306 (1964).
36. J. Munoz and R. K. Bergman, *Bacteriol. Rev.*, *32*, 103 (1968).
37. R. K. Bergman and J. Munoz, in press.
38. J. Bordet and Sleeswyk, *Ann. Inst. Pasteur*, *24*, 476 (1910).
39. L. F. Schuchardt, J. Munoz, W. F. Verwey, and J. F. Sagin, *J. Immunol.*, *91*, 107 (1963).

REFERENCES

40. E. K. Andersen, *Acta Pathol. Microbiol. Scand.*, *33*, 202 (1953).
41. G. Eldering, C. Hornbeck, and J. Baker, *J. Bacteriol.*, *74*, 133 (1957).
42. G. Eldering, W. C. Eveland, and P. L. Kendrick, *J. Bacteriol.*, *83*, 745 (1962).
43. G. Eldering, Round Table Conference on Pertussis Immunization, Prague, *1*, 81 (1962).
44. R. Ross and J. Munoz, *Infect. Immun.*, *3*, 243 (1971).
45. J. J. Cameron, *Pathol. Bacteriol.*, *94*, 367 (1967).
46. T. N. Stanbridge and N. W. Preston, *J. Hyg. Camb.*, *73*, 305 (1974).
47. L. B. Holt, *J. Med. Microbiol.*, *1*, 169 (1968).
48. E. W. Flosdorf and A. C. Kimball, *J. Immunol.*, *39*, 475 (1940).
49. J. Smolens and S. Mudd, *J. Immunol.*, *47*, 155 (1943).
50. H. Schweinberg, *Behringwerk-Mitt.*, *40*, 102 (1961).
51. K. Onoue, M. Kitagawa, and Y. Yamamura, *J. Bacteriol.*, *82*, 648 (1961).
52. E. W. Flosdorf, H. M. Felton, A. Bondi, Jr., and A. C. McGuinness, *Amer. J. Med. Sci.*, *206*, 422 (1943).
53. M. Pittman, *J. Wash. Acad. Sci.*, *46*, 234 (1956).
54. P. L. Kendrick, G. Eldering, M. K. Dixon, and J. Misner, *Amer. J. Public Health*, *37*, 803 (1947).
55. M. Pittman and J. E. Lieberman, *Amer. J. Public Health*, *38*, 15 (1948).
56. Medical Research Council, Vaccination Against Whooping Cough: Final Report, *Brit. Med. J.*, *1*, 994 (1959).
57. G. Eldering, J. Holwerda, and J. Baker, *J. Bacteriol.*, *91*, 1759 (1966).
58. J. M. Dolby and C. J. Bronne-Shanbury, *J. Biol. Standard.*, *3*, 89 (1975).
59. G. Eldering, J. Holwerda, and J. Baker, *J. Bacteriol.*, *93*, 1758 (1966).
60. R. K. Bergman and J. Munoz, *J. Allergy Appl. Immunol.*, *55*, 378 (1975).
61. N. Preston, *J. Pathol. Bacteriol.*, *91*, 173 (1966).
62. N. W. Preston and T. N. Stanbridge, *Brit. Med. J.*, *19*, 448 (1972).
63. N. W. Preston, *Brit. Med. J.*, *2*, 724 (1963).
64. N. Chalvardjian, *Can. Med. Assoc. J.*, *92*, 1114 (1965).
65. A. C. Blaskett, J. Gulasekharam, and L. C. Fulton, *Med. J. Australia*, *1*, 781 (1971).

66. G. Eldering, J. Holwerda, A. Davis, and J. Baker, *Appl. Microbiol.*, *18*, 618 (1969).
67. T. N. Stanbridge and N. W. Preston, *J. Hyg. Camb.*, *72*, 213 (1974).
68. E. V. Keogh, E. A. North, and M. F. Warburton, *Nature*, *160*, 63 (1947).
69. E. V. Keogh and E. A. North, *Australian J. Exptl. Biol. Med. Sci.*, *26*, 315 (1948).
70. F. L. G. Masry, *J. Gen. Microbiol.*, *7*, 201 (1952).
71. S. Fisher, *Australian J. Exptl. Biol. Med. Sci.*, *28*, 509 (1950).
72. Y. Watanabe, *Kitasato Arch. Exptl. Med.*, *26*, 151 (1953).
73. E. H. Thiele, *J. Immunol.*, *65*, 627 (1950).
74. S. Fisher, *Brit. J. Exptl. Pathol.*, *29*, 357 (1948).
75. L. Pillemer, *Proc. Soc. Exptl. Biol. Med.*, *75*, 704 (1950).
76. Y. Sato and H. Arai, *Infect. Immun.*, *6*, 899 (1972).
77. A. Adler and S. I. Morse, *Infect. Immun.*, *7*, 461 (1973).
78. S. I. Morse and J. H. Morse, *Fed. Proc.*, *33*, 763 (1974).
79. S. I. Morse and J. H. Morse, *J. Exptl. Med.*, *143*, 1483 (1976).
80. J. Bordet and O. Gengou, *Ann. Inst. Pasteur*, *20*, 731 (1906).
81. H. Felton and C. Y. Willard, *J. Amer. Med. Assoc.*, *126*, 294 (1944).
82. L. Sauer, *J. Amer. Med. Assoc.*, *101*, 1449 (1933).
83. P. Kendrick and G. Eldering, *Amer. J. Public Health*, *25*, 147 (1935).
84. H. M. Felton and W. F. Verwey, *Pediatrics*, *6*, 637 (1955).
85. L. Pillemer, L. Blum, and I. H. Lepow, *Lancet*, *1*, 1257 (1954).
86. J. Munoz and B. M. Hestekin, *Proc. Soc. Exptl. Biol. Med.*, *112*, 799 (1963).
87. M. Niwa, *J. Biochem. (Tokyo)*, *51*, 222 (1962).
88. R. E. Pieroni, E. J. Broderick, and L. Levine, *J. Immunol.*, *95*, 643 (1965).
89. Y. Sato, H. Arai, and K. Suzuki, *Infect. Immun.*, *9*, 801 (1974).
90. S. B. Lehrer, E. M. Tan, and J. H. Vaughan, *J. Immunol.*, *113*, 18 (1974).
91. C. W. Parker and S. I. Morse, *J. Exptl. Med.*, *137*, 1078 (1973).
92. Food and Drug Administration, Title 21, Code of Federal Regulations, Part 600, Biological Products, U.S. Dept. Health, Education and Welfare, Washington, D.C., U.S. Printing Office (1975).

REFERENCES

93. World Health Organization, Technical Report Series, 1964, No. 274.
94. P. van Hemert and H. Cohen, *Round Table Conference on Pertussis Immunization, Prague, 1,* 24 (1962).
95. D. W. Stainer, *Symp. Series Immunobiol. Standard.,* Vol. 13, Karger, Basel, 1970, p. 89.
96. I. Joo, *Symp. Series Immunobiol. Standard.,* Vol. 13, Karger, Basel, 1970, p. 48.
97. J. Munoz and B. M. Hestekin, *J. Immunol., 91,* 2175 (1966).
98. L. Greenberg and D. S. Fleming, *Can. J. Public Health, 39,* 131 (1948).
99. D. S. Fleming, L. Greenberg, and E. M. Beith, *Can. Med. Assoc. J., 59,* 101 (1948).
100. I. A. Parfentjev and M. A. Goodline, *J. Pharmacol. Exptl. Therap., 92,* 411 (1948).
101. M. Pittman, *J. Infect. Diseases, 89,* 296 (1951).
102. L. S. Kind and R. H. Gadsden, *Proc. Soc. Exptl. Biol. Med., 84,* 373 (1953).
103. J. Munoz and L. F. Schuchardt, *J. Allergy, 24,* 330 (1953).
104. R. S. Abernathy and W. W. Spink, *J. Immunol., 77,* 418 (1956).
105. L. S. Kind, *Proc. Soc. Exptl. Biol. Med., 95,* 200 (1957).
106. P. Kallos and L. Kallos-Deffner, *Arch. Allergy Appl. Immunol., 11,* 237 (1957).
107. J. Munoz, *Proc. Soc. Exptl. Biol. Med., 95,* 328 (1957).
108. M. Pittman and F. C. Germuth, *Proc. Soc. Exptl. Biol. Med., 87,* 425 (1954).
109. J. Munoz and R. L. Anacker, *J. Immunol., 83,* 502 (1959).
110. I. A. Parfentjev, *Yale J. Biol. Med., 27,* 46 (1954).
111. Y. Sato, H. Arai, and K. Suzuki, *Infect. Immun., 7,* 992 (1973).
112. S. Levine, E. J. Wenk, H. B. Devlin, R. E. Pieroni, and L. Levine, *J. Immunol., 97,* 363 (1966).
113. J. Munoz, R. F. Smith, and R. L. Cole, International Symposium on Pertussis, Biltnoven, 1969, *Symp. Series Immunobiol. Standard.,* Vol. 13, Karger, Basel, 1970, p. 205.
114. S. B. Lehrer, J. H. Vaughan, and E. M. Tan, *J. Immunol., 114,* 34 (1975).
115. A. C. Wardlaw and C. M. Jakus, *Can. J. Microbiol., 12,* 1105 (1966).
116. A. C. Wardlaw, *Symp. Series Immunobiol. Standard.,* Vol. 3, Karger, Basel, 1967, p. 99.

117. T. Hiramatsu, *J. Osaka City Med. Center, 16,* 693 (1960).
118. S. Iwasa, S. Ishida, S. Asakawa, and M. Kurokawa, *Jap. J. Med. Sci. Biol., 21,* 363 (1968).
119. S. I. Morse and K. K. Bray, *J. Exptl. Med., 129,* 523 (1969).
120 S. B. Lehrer, R. M. Nakamura, and E. M. Tan, *Int. Arch. Allergy Appl. Immunol.,* in press (1976).
121. H. Svensson, in *A Laboratory Manual of Analytical Methods in Protein Chemistry Including Polypeptides* (P. Alexander and R. J. Bloch, eds.), Vol. 1, Pergamon Press Ltd., London, 1960, pp. 193-244.

Chapter 3

SHOCK-ENHANCING EFFECTS OF PERTUSSIGEN

I.	DESCRIPTION OF PHENOMENON.	72
	A. Characteristics of Circulatory Shock.	72
	B. Early Observations on Shock Enhancement by *B. pertussis*.	72
II.	FACTORS THAT AFFECT THE ENHANCEMENT OF SHOCK BY PERTUSSIGEN.	73
	A. Animal Species.	73
	B. Mouse Strains	76
	C. Age of Mice	76
	D. Sex of Mice	77
	E. Environmental Factors and Stress.	77
	F. Physical Form of Pertussigen and Route of Administration.	81
	G. Shock-Inducing Treatments	82
	H. Dose of Pertussigen	82
III.	HYPOTHESES ABOUT SHOCK ENHANCEMENT BY PERTUSSIGEN.	87
	A. Early Hypotheses.	87
	B. Characteristics of Histamine and Anaphylactic Shock in the Mouse.	89
	C. Adrenergic Involvement.	93
	D. Summary	102
	REFERENCES	103

I. DESCRIPTION OF PHENOMENON

A. Characteristics of Circulatory Shock

The complex phenomenon of circulatory shock is not brought on by a single etiological factor, but results from different pathological events. However, the one common feature of different forms of shock is inadequate tissue perfusion. Three general classifications of shock have been described: (a) hypovolemic shock resulting when a significant loss of blood or plasma from the vascular bed occurs due to hemorrhage or greatly increased vascular permeability, (b) cardiogenic shock due to inadequate pumping action of the heart to maintain circulation, and (c) low-resistance shock due to a generalized vasodilatation which in spite of little or no change in cardiac output and blood volume results in greatly diminished blood circulation (1).

B. Early Observations on Shock Enhancement by *B. pertussis*

Eldering (2) found that administration of a polysaccharide preparation from *B. pertussis* cells into mice sensitized them to a subsequent infection with live *B. pertussis* cells. Since most of the mice died within 2 h after receiving the live cells the cause of death was probably anaphylactic shock, and thus it seems to be the first recorded observation of an increased susceptibility to shock induced by a substance from *B. pertussis*. Soon after this observation, Ospeck and Roberts (3) reported that mice which had been treated with a crude toxoid preparation, made from culture filtrates of *B. pertussis,* often died of shock when challenged with the toxin. This clearly showed that *B. pertussis* had the ability to enhance anaphylactic shock, although it was not recognized as such by the authors. In 1947, Parfentjev et al. (4-6) published their observations that mice treated with pertussis vaccine died after receiving a denatured nucleoprotein isolated from *B. pertussis*. This hypersensitivity was thought to have some immunological basis, but this phenomenon was not clearly associated with anaphylactic shock until

II. FACTORS THAT AFFECT SHOCK ENHANCEMENT

Malkiel and Hargis (7-10) found that *B. pertussis* treatment greatly enhanced anaphylactic shock in mice inoculated with other antigens. Their observations were confirmed and extended by other workers (11) using a variety of antigens; furthermore, *B. pertussis*-treated mice were found to be more susceptible to passively induced anaphylaxis (12-15). Following their observations (4-6) on the sensitivity of *B. pertussis*-treated mice to the denatured protein of *B. pertussis*, Parfentjev and Goodline tested the histamine sensitivity of *B. pertussis*-treated mice. They found that treatment with pertussis vaccine increased histamine sensitivity up to 100-fold (16). Their work was soon confirmed (11) and later it was discovered that certain mouse strains would also become hypersensitive to serotonin when treated with pertussis vaccine (17-20). This observation was felt to be quite significant because serotonin had been implicated in mouse anaphylaxis (21-23). Subsequent studies have shown that *B. pertussis*-treated mice are much more susceptible to several agents or treatments that affect the vascular bed. For example, increased sensitivity has been reported for all of the following treatments in *B. pertussis*-treated mice: bacterial infections (24-28), endotoxins (29-35), x-irradiation (18,36,37), anoxia (38), cold stress (39), peptone shock (36,40-42), bradykinin (43), and methacoline (44). From this list it appears that in the mouse, *B. pertussis*-treatment alters or inhibits a basic physiological function which normally acts to protect against the toxic effects of many substances that kill the animal through hypovolemic or low-resistance shock.

II. FACTORS THAT AFFECT THE ENHANCEMENT OF SHOCK BY PERTUSSIGEN

A. Animal Species

Many factors block or modify the shock-enhancing properties of pertussigen. Its effectiveness depends on the particular species of animal in which it is tested and furthermore upon the strain within a species. In addition to the mouse, the rat (10) and the chicken (45) can be sensitized to histamine. Species, such as the guinea pig and rabbit, which have a very high natural sensitivity to histamine, do not become more sensitive after pertussigen treatment and in fact may become more resistant (46,47). Hamsters did not develop any histamine sensitivity after pertussigen treatment, while bats may have developed some slight sensitivity (R. K. Bergman, unpublished observations).

B. Mouse Strains

Among mouse strains, there are marked differences in susceptibility to histamine sensitization after pertussigen treatment. Several strains, mainly those derived from the Swiss-Webster line, are highly responsive and 50- to 100-fold increases in sensitivity are common. Several other strains including several inbred strains are quite resistant to histamine sensitization. Some strains develop serotonin sensitivity without histamine sensitivity. A number of strains (inbred and noninbred) did not sensitize to histamine, and in some cases also not to serotonin, but they did become sensitive to a combined challenge of histamine and serotonin (48), see Table 1, p. 76. Probably most mouse strains are affected to some degree by pertussigen, but their susceptibility to shock is dependent upon the severity of the challenge treatment.

C. Age of Mice

Age, especially in certain mouse strains, can affect the shock-inducing effects of pertussigen (52,55,59). A strain reared at the Rocky Mountain Laboratory (RML) was not very susceptible to the effects of pertussigen at 3 to 7 weeks of age, but mice over 7 weeks old did become sensitive to histamine shock after pertussigen treatment (55). This may be related to the fact that in some strains there is a natural increase in histamine sensitivity of normal mice as they become older (58). On the other hand, age in certain strains, such as the CFW, does not seem as important, although from our experience more uniform sensitivity is observed in 5- to 7-week-old mice than in younger CFW mice.

D. Sex of Mice

Some workers have found that the sex of mice affects their susceptibility to shock after pertussigen treatment. Kind (59) did not find a difference in responses between sexes in individual experiments, but we, along with Pittman (62) and Maitland et al. (46), found that males were less responsive in a large number of experiments when evaluated statistically. This difference in an individual experiment may not be obvious, and may even be reversed. In some cases we suspect that the lower susceptibility in males may result from stress produced by fighting. In experiments employing male mice, it is best not to mix groups in which hierarchical orders have become established.

E. Environmental Factors and Stress

Environmental factors such as housing conditions and diet alter the responsiveness of mice to the shock-enhancing effect of pertussigen. Mice housed individually were not as susceptible as were mice caged in groups of 10 (63). In mice caged individually, those receiving a purified diet were less sensitive (J. J. Munoz and D. Boshart, unpublished observations)(Table 2) than those fed a commercial pellet ration. Substances which may produce mild stress or low-grade irritation also seem to inhibit or decrease the effect of pertussigen. For example, toxic preparations of *B. pertussis* cells produce weaker histamine sensitivity than do cell preparations which have been detoxified by heating at 55°C for 30 min (64). Intraperitoneal injection of aspirin, propylene glycol, and other substances to mice during a 4-day interval between giving pertussigen and histamine challenge decreases the sensitivity of the mice (65). Curiously, the severity of histamine shock does not follow a straight line dose-dependent relationship. Frequently, higher mortality is obtained with 0.5 to 2 mg histamine challenge than with a higher dose (4 to 8 mg). The phenomenon is enhanced by prior treatment with aspirin or propylene glycol (65)(Fig. 1).

TABLE 1

Susceptibility of Mouse Strains to Sensitizing Effects of Pertussigen[a]

Strain	Breeder	Sensitization to:[b]			References
		Histamine	Serotonin	Serotonin + histamine	
Noninbred					
CFW[c]	Carworth Farms	+	+		49, 50, 51
CF-1[d]	Carworth Farms	−	+		17, 42, 52
BF[b]	Beverly Farms	+			52
TF[b]	Tumblebrook Farms	+	+		19, 53, 54
SD$_1$[c]	Sharp and Dohme	+			52
SD$_2$	Sharp and Dohme	−			52
RML (young)	RML	−	−	+	55, 56
RML (old)	RML	+			55
NIH[c]	NIH	+			48
GP[c]	NIH	+		+	48
Swiss-Webster	Taconic Farms	+			57
NLA[c]	Natl. Lab. Animal Co.	+	+		19
Prinston	Millerton Farm	−			50
NIH-BS[c]	NIH	+			47
C58, F-1	Millerton Farm	−			58
Inbred					
AAF$_1$	NIH	−		+	48
CDF$_1$	NIH	−		+	48

II. FACTORS THAT AFFECT SHOCK ENHANCEMENT

Strain	Source				Ref.
CAF$_1$	NIH		−	+	48
C3H^{fb}/HeN	NIH		−	+	48
AL/N	NIH		−	+	48
C57L/N	NIH		+	+	48
STR/1N	NIH	+	−	+	48
BRSUNT/N	NIH		−	+	48
C57BL/10ScN	NIH		−	+	48
DBA/2N	NIH	−	−	+	48
C57BL/6N	NIH	−	−	+	48
Balb/c AnN	NIH	+	−	+	48
C3H/HoN	NIH	−	−	−	48
A/HeN	NIH	−	−	+	48
STR/N	NIH	−			48
AKR/N	NIH	+−	+−		48
AKR	Millerton Farms	+			50
Isko	Not given	+	+		59
LAF$_1$	Not given	+−	+		60

[a] Reprinted from Ref. 11, p. 109, by courtesy of the American Society for Microbiology.
[b] Symbols: + = susceptible to sensitization; − = not susceptible to sensitization; +− = questionable sensitization; blank = not tested.
[c] Known to originate from Swiss-Webster strain.
[d] A substrain of CF$_1$ raised in the University of Montreal has been found susceptible to histamine sensitization by Guerault (61).

TABLE 2

Effects of Group vs. Individual Caging and Diet on Histamine Sensitivity of Pertussigen-Treated Mice[a]

Type of caging	Type of diet[b]		Totals
	Purified	Purina	
Grouped	70/76 (92%)[c]	74/79 (94%)	144/155 (93%)
Individual	41/79 (52%)	57/79 (72%)	98/158 (62%)
Totals	111/155 (72%)	131/158 (83%)	

[a] Tumblebrook Farms ♀♀ 10 to 12 g body weight.
[b] Half of mice were fed a purified diet and half were fed Purina Laboratory Chow.
[c] Number of mice dead/number mice challenged with histamine (% mortality). Grouped mice were significantly more sensitive than individually caged mice ($P \ll 0.001$). Mice on purified diet were significantly less sensitive than those on Purina diet ($P < 0.02$). Data from J. J. Munoz and D. Boshardt, unpublished observations.

We also noted that the normal resistance of CFW mice to histamine can be diminished by a number of environmental factors which are not very well understood or characterized. Several years ago, we noted a striking variability in the histamine sensitivity of normal mice 4 days after their arrival in our laboratory, following a 2,000 mile trip by air express, which required the animals to be in transit for 36 to 48 h. As shown in Fig. 2, mortality rate following histamine challenge frequently fell between 30 to 50% in these mice. Usually this histamine sensitivity decreased within 1 to 2 weeks to a tolerable level so that the mice could be used in our experimental work.

Because mice were very thirsty after 36 to 48 h in transit, we initiated experiments to determine if dehydration could make CFW mice hypersensitive to histamine. Table 3 summarizes the results of an experiment in which groups of 20 mice were maintained without water for 1 to 4 days. After each respective dehydration period, 10 mice were challenged i.p. with 0.5 mg histamine. The remaining 10 mice were allowed to have water ad lib for 3 days and then chal-

II. FACTORS THAT AFFECT SHOCK ENHANCEMENT

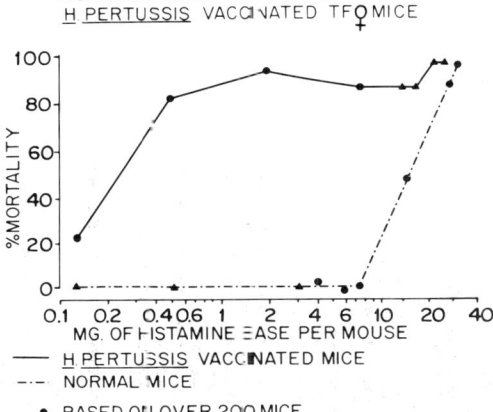

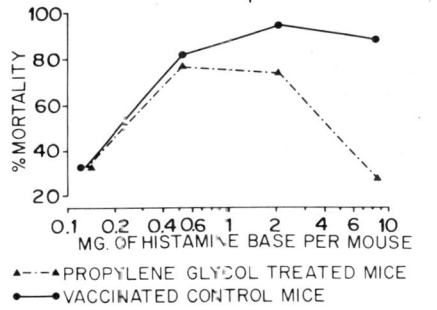

FIG. 1. Biphasic response of pertussigen-treated mice (2 x 10^9 killed *B. pertussis* cells were given i.p. 4 days before challenge) to increasing doses of histamine. Mortality actually decreases when mice are challenged with 4 to 8 mg of histamine base and this phenomenon is enhanced by treating mice with propylene goycol. (Reprinted from Ref. 65, p. 121, by courtesy of the C. V. Mosby Co.)

lenged with histamine. Dehydration alone did not cause mice to become sensitive to histamine. However, mice which were without water for 3 to 4 days and then allowed to drink for 3 days did become quite sensitive. Onset and duration of histamine sensitivity in mice which were allowed water ad lib after a 4-day dehydration period, as shown in Fig. 3, was quite transient. Sensitivity began to increase

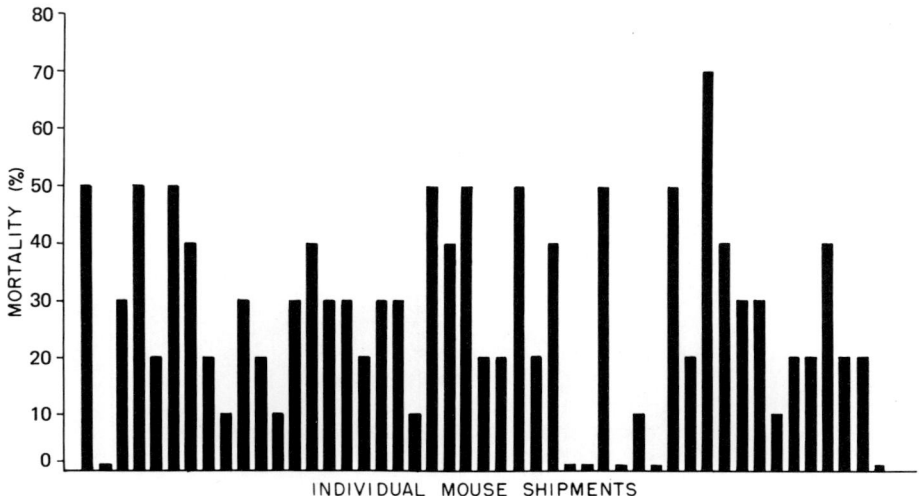

FIG. 2. Variation in sensitivity to 0.5 mg of histamine given i.p. to different shipments of normal, untreated CFW female mice received at the Rocky Mountain Laboratory from New York. Four days after arrival, animals were challenged in groups of 10. Testing covered a period of 11 months from March 1967 through January 1968. No correlation between season of year and mortality was evident. (Reprinted from Ref. 11, p. 111, by courtesy of the American Society for Microbiology.)

TABLE 3

Effect of Dehydration and Subsequent Rehydration on Histamine Sensitivity in CFW Mice[a,b]

Group	Days of dehydration				3 days after rehydration
	1	2	3	4	
1	3/10[c]				3/10
2		1/10			2/10
3			0/10		6/10
4				0/10	7/10

[a] Data from R. K. Bergman (unpublished observations).
[b] Thirty mice from the same lot were housed with food and water ad lib and challenged on days 0, (3/10), 3 days (0/10), and 7 days (2/10).
[c] Deaths/total mice challenged (0.5 mg histamine given i.p.).

II. FACTORS THAT AFFECT SHOCK ENHANCEMENT

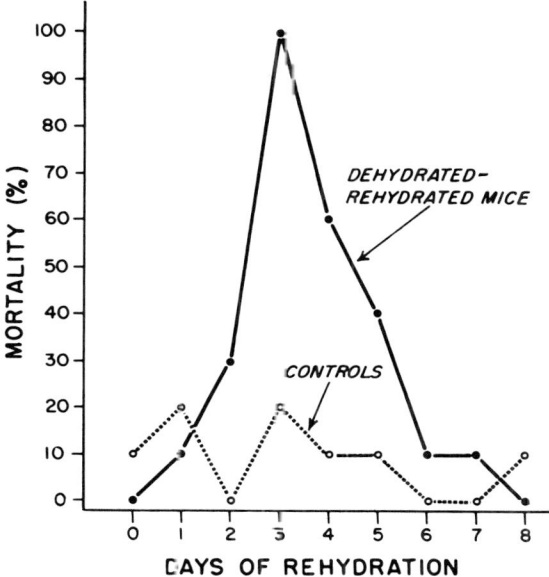

FIG. 3. Onset and duration of histamine sensitivity of rehydrated mice after a 4 day period of dehydration. Mice were challenged i.p. with 0.5 mg of histamine base. Control mice were allowed water ad lib throughout the experiment. (Taken from R. K. Bergman, unpublished observations.)

on the 2nd day of rehydration, reached its maximum on the 3rd day, and then returned to normal on the 6th day. We cannot readily explain the phenomenon, but we think the sympatho-adrenal medullary system is less responsive to a secondary stress closely following an initial stress.

F. Physical Form of Pertussigen and Route of Administration

The effectiveness of pertussigen in enhancing shock depends on the physical form in which it is given and upon the route of administration. In general, soluble preparations made from cell lysates are more effective than whole cell preparations. Whole cell preparations are most effective when administered i.v. and least potent when given s.c. (46,63,66,67; J. J. Munoz and R. K. Bergman, unpublished observations)(Fig. 4). Soluble preparations are not so dependent on route

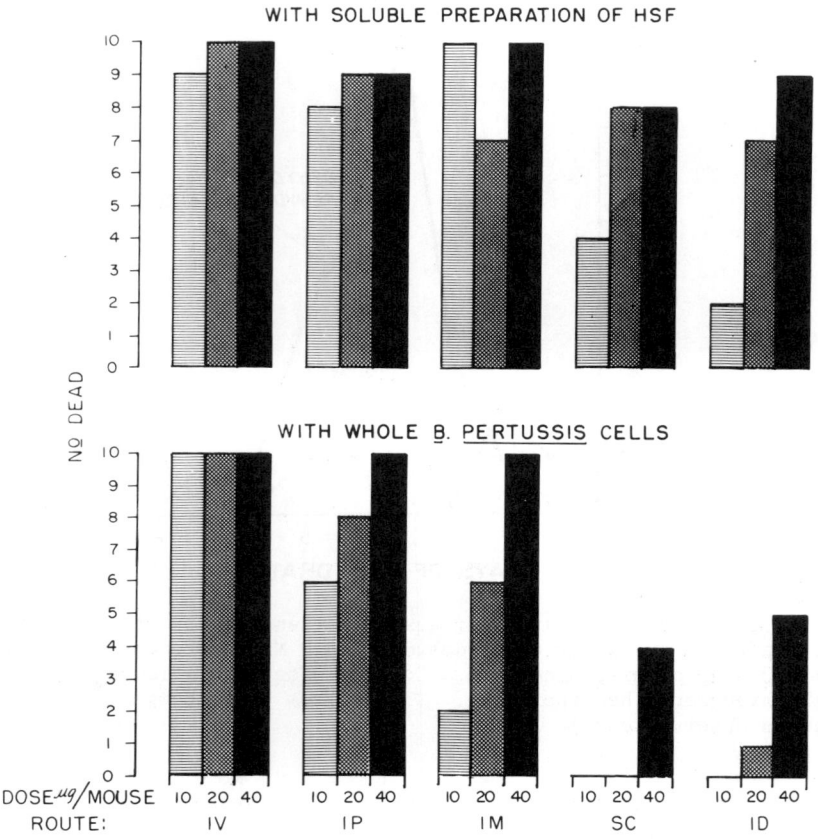

FIG. 4. Effect of route of administration and type of HSF preparation on histamine sensitization of CFW female mice. The doses of sensitizing material in the upper part of the figure correspond to those indicated in the lower portion. IV - intravenous sensitization; IP = intraperitoneal sensitization; IM = intramuscular sensitization; SC = subcutaneous sensitization; ID = intradermal sensitization. (Reprinted from Ref. 11, p. 112, by courtesy of the American Society for Microbiology.)

of administration, but even so, the i.v. route is best (Fig. 5). The i.c. injection of whole cell vaccine (68) or the intranasal inoculation with live cells (69) have also been reported to be effective routes to increase the sensitivity of mice to histamine shock. Geller and Pittman (70) demonstrated that with live *B. pertussis* the

II. FACTORS THAT AFFECT SHOCK ENHANCEMENT

intranasal route was more effective than the i.p. route and that very small doses of pertussis vaccine induced a late onset of hypersensitivity to histamine.

G. Shock-Inducing Treatments

Much of the foregoing discussion on factors which regulate or influence shock enhancement by pertussigen pertains primarily to shock induced by histamine. Probably many of these factors also affect serotonin and endotoxin shock. However, the susceptibility of mice to various forms of shock do not exactly parallel each other following administration of pertussigen. As can be seen in Fig. 6, sensitivity to histamine following i.v. administration of soluble pertussigen rises very quickly and lasts many days. Histamine sensitivity following an i.p. injection of *B. pertussis* cells rises more slowly, reaching a peak in 4 or 5 days and then returns to normal in about 32 days; serotonin sensitivity follows a somewhat similar time course. Endotoxin sensitivity is somewhat slower in its onset, not becoming maximal until about 12 days, but then lasting beyond 32 days.

Figure 7 demonstrates the striking effect of pertussigen treatment on histamine sensitivity in mice. When histamine LD_{50} values were calculated for normal and treated mice it was obvious that pertussigen was still affecting histamine sensitivity at least up to 84 days (2,016 h) postinoculation.

H. Dose of Pertussigen

If sensitivity is quantitated by calculating histamine LD_{50} values for groups of mice which have received graded doses of pertussigen the range of pertussigen doses which produce a linear dose response is quite narrow. (Fig. 8). A dose of 0.156 µg SE produced little if any sensitization and 2.5 µg made the mice almost maximally sensitive. The greatest change in sensitivity was produced by doses between 0.312 and 1.25 µg. These data suggest that there is a limited number of sites in the mouse which are affected by SE, and once these are covered or blocked by the active material the animal is maximally sensitized.

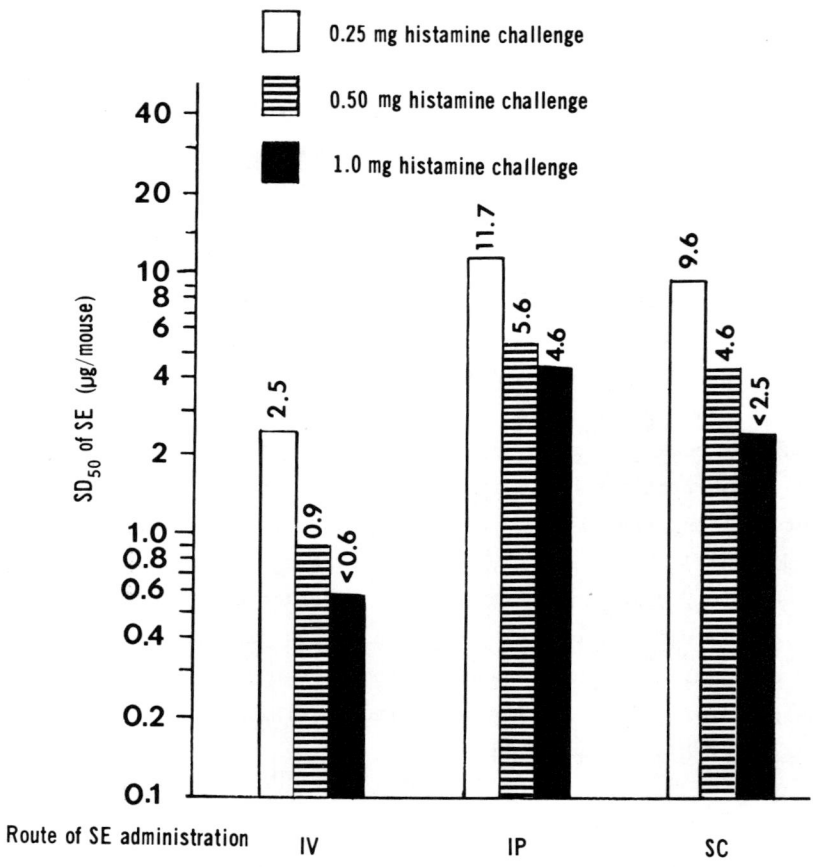

FIG. 5. Sensitizing dose$_{50}$ (SD)$_{50}$ of saline extract (SE) from *B. pertussis* administered to mice by i.v., i.p., and s.c. routes as determined by challenging mice 4 days later i.p. with 3 different doses of histamine. (Data from Ref. 67.)

II. FACTORS THAT AFFECT SHOCK ENHANCEMENT

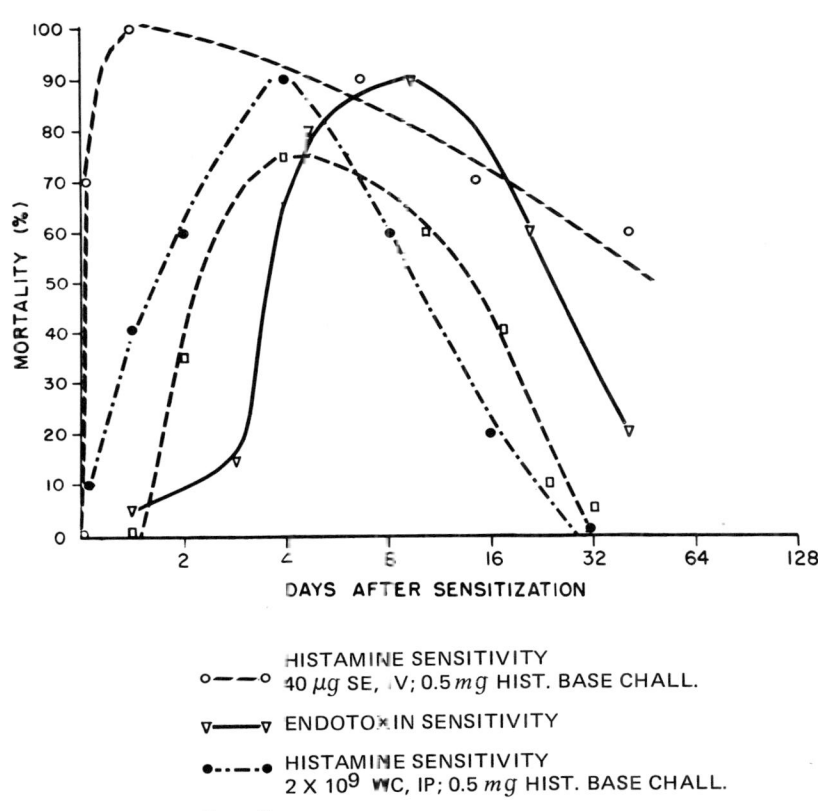

FIG. 6. Time course of sensitivity to histamine, serotonin, and endotoxin in mice treated with pertussigen. The endotoxin and serotonin sensitivity was measured in mice treated with whole cell vaccines. (Data taken from Ref. 11.)

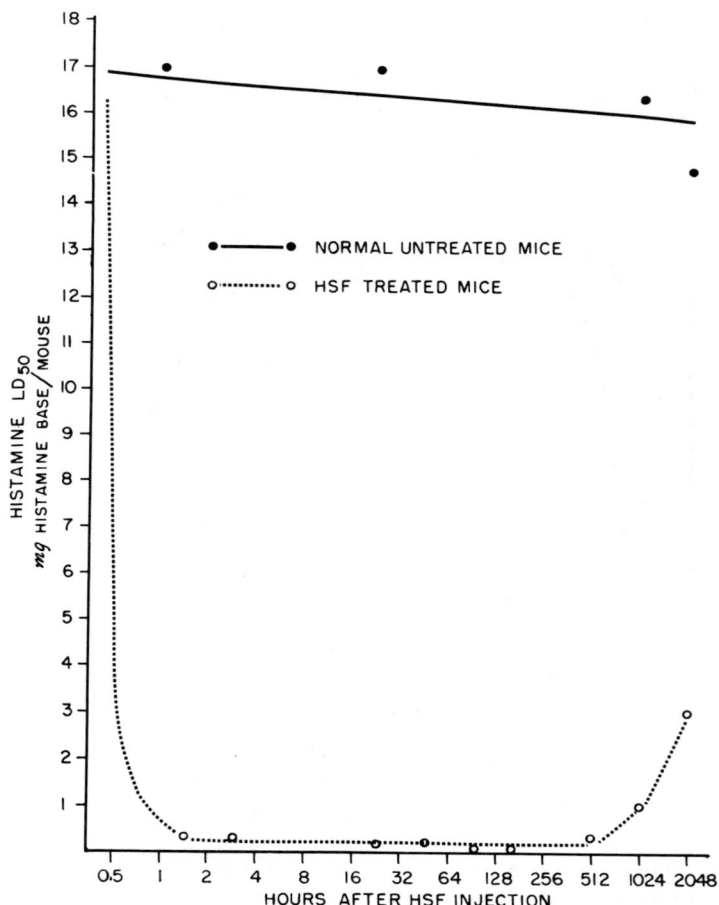

FIG. 7. Onset and pertistence of histamine sensitivity in CFW mice treated with alkaline-saline extract of *B. pertussis* given i.v. (Reprinted from Ref. 11, p. 110, by courtesy of the American Society for Microbiology.)

III. HYPOTHESES ABOUT SHOCK ENHANCEMENT 87

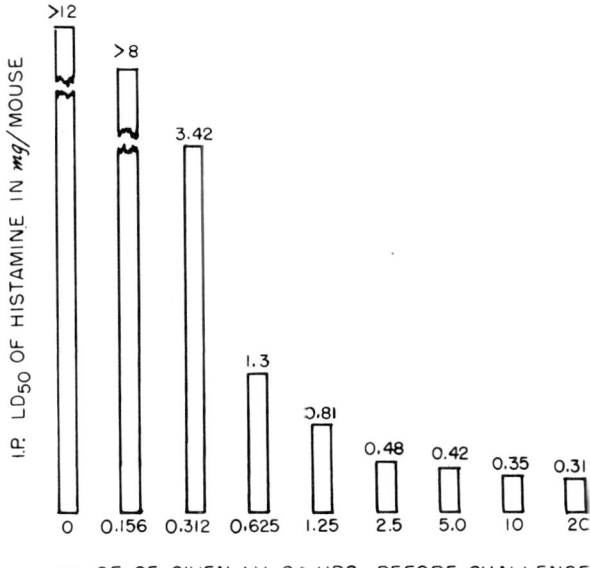

FIG. 8. Effect of amount of alkaline saline extracts (SE) on LD_{50} of histamine. (Reprinted from Ref. 67, p. 123, by courtesy of Williams and Wilkins Co., copyright 1966.)

III. HYPOTHESES ABOUT SHOCK ENHANCEMENT BY PERTUSSIGEN

A. Early Hypotheses

The physiological changes by which pertussigen enhances susceptibility to shock are not yet fully understood. A number of hypotheses have been proposed, but none explain all of the experimental observations. Some early hypotheses proposed that pertussigen altered the rate at which endogenous histamine was destroyed or produced in susceptible animals. A number of investigators reported that histaminase activity of tissues was reduced in animals treated with pertussis vaccine (71-77). However, this observation did not provide a satisfactory answer for a number of reasons: (a) guinea pigs treated with pertussis vaccine develop a lower tissue histaminase level, and yet they do not demonstrate a greater histamine sensitivity, (b) hyper-sensitivity to other treatments (serotonin, peptone, anaphylaxis, cold stress, x-rays, etc.) could not be explained by depressed tissue histaminase levels, and (c) in the mouse, the main histamine detoxifying system is via methylation and oxidation and not histaminase (78).

While tissue histaminase was found to be depressed, histidine decarboxylase, which decarboxylates histidine to form histamine, was found to be elevated in animals receiving pertussis vaccine (79). It was suggested that this would increase the levels of endogenous histamine in the tissues and make the animals more sensitive to exogenous histamine. Since it was also found that histidine decarboxylase was equally elevated in mouse strains that do not become histamine sensitive (79) and that endotoxin from Gram-negative bacteria produced elevated histidine decarboxylase levels without concomitant histamine sensitivity (79,80), the importance of this enzyme in histamine shock enhancement was greatly diminished. This hypothesis was conclusively nullified when Szentivanyi et al. reported that pertussigen preparations free of endotoxin do not increase histidine decarboxylase levels (81).

The metabolism of certain substances is markedly altered in pertussigen-treated mice and rats (Table 4). At least part of these altered metabolic responses may be due to the ability of pertussigen to induce hyperinsulinemia in mice and rats (82,83). The response of these rodents to the hyperinsulinemia is somewhat species dependent. Blood glucose is markedly depressed in mice (47,84-87) but is only slightly depressed in rats (82). Epinephrine-induced hyperglycemia in pertussigen-treated mice was blocked (87), but in rats it is only slightly reduced (82). Pertussigen treatment of mice blocks epinephrine-induced free fatty acidemia (88) but is not effective in rats (82). Some investigators thought that mouse sensitivity to histamine and other agents might be realted to the hypoglycemic effect (89), but 4 observations cast doubt on this hypothesis: (a) insulin in small doses, which produced hypoglycemia equal to or greater than pertussigen treatment, failed to produce histamine hypersensitivity (90), (b) large doses of glucose or other monosaccharides failed to protect pertussigen-treated mice from histamine

III. HYPOTHESES ABOUT SHOCK ENHANCEMENT

TABLE 4

Effects of Pertussigen on Glucose and Fatty Acid Metabolism in Mice and Rats

Parameter	Response
Blood glucose level	Markedly depressed in mice
	Slightly depressed in rats
Epinephrine-induced hyperglycemia	Blocked in mice
	Slightly depressed in rats
Epinephrine-induced lactic acidemia	Blocked in mice
Epinephrine-induced free fatty acidemia	Blocked in mice
	Not affected in rats
Glucose tolerance	Greatly increased in mice

shock (84,90), (c) a short-term diabetogenic agent, D-mannoheptulose, administered to pertussigen-treated mice did not protect them from histamine sensitivity (90), and (d) the time course of hypoglycemia in pertussigen-treated mice does not correspond to the period when mice are susceptible to shock (90). Cronholm and Fishel (91) also noted a lack of relationship between hypyglycemia and histamine sensitivity.

A role for the reticuloendothelial system has been postulated for enhanced shock in pertussigen-treated mice. "Blockade" of this system with colloidal particles protects mice from anaphylaxis (92) and pertussigen-treated mice from the effects of serotonin (93). The manner by which this protection occurs is not clear mainly because this problem has not been thoroughly investigated.

B. Characteristics of Histamine and Anaphylactic Shock in the Mouse

Before presenting more recent hypotheses about the shock-enhancing effects of pertussigen, it is germane to discuss the salient features of shock in the mouse. There has been some confusion as to what the exact cause of death and the "shock organ" are in mouse anaphylaxis and histamine shock. In the mouse, respiratory distress is seen in anaphylactic shock or from histamine challenge, but they do not undergo the bronchial spasm and asphyxia like that seen in anaphylaxis in guinea pigs. They do not respond like the rabbit in anaphylactic shock, in which acute constriction of the pulmonary arterioles and extreme dilatation of the right heart results in acute heart failure. Nor are mice similar to the dog in anaphylactic shock where death follows severe liver congestion. A mouse examined immediately after anaphylactic death has collapsed and unobstructed lungs, a heart that is still beating and an uncongested liver. Except for a generalized injected and edematous appearance in the intestines and stomach, the abdominal organs appear normal. McMaster and Kruse described events following initiation of anaphylactic shock in mice (94). First, there was an intense spasm of the arterioles and venules within 1 min after injecting antigen into sensitized animals. Within the next 5 to 20 min there was a marked relaxation of the arterioles and venules and a fall in blood pressure and all the vessels became dilated with slow-moving blood. A number of investigators have reported that there is a marked rise in hematocrit in mice undergoing anaphylactic shock, indicating a loss of fluid volume from the blood (95-97)(Fig. 9). We estimated from the elevation in hematocrit values that mice undergoing anaphylaxis or injected with 0.5 to 1.0 mg histamine lost 26 to 34% of their blood volume. It must also be pointed out that normal mice, i.e., mice not treated with pertussigen, also experience marked hemoconcentration in anaphylaxis or after receiving histamine, but they ordinarily survive. The singular importance of the vascular bed in mouse anaphylaxis or in response to histamine shock was confirmed when it was found that these mice could be protected from death by i.v. administration of 6% dextran solution or physiological saline as blood volume expanders (97)(Table 5). Epinephrine is highly effective for treating mice in histamine shock. It was most effective when given immediately after histamine challenge, and its effectiveness was diminished markedly if the mice had received large

III. HYPOTHESES ABOUT SHOCK ENHANCEMENT

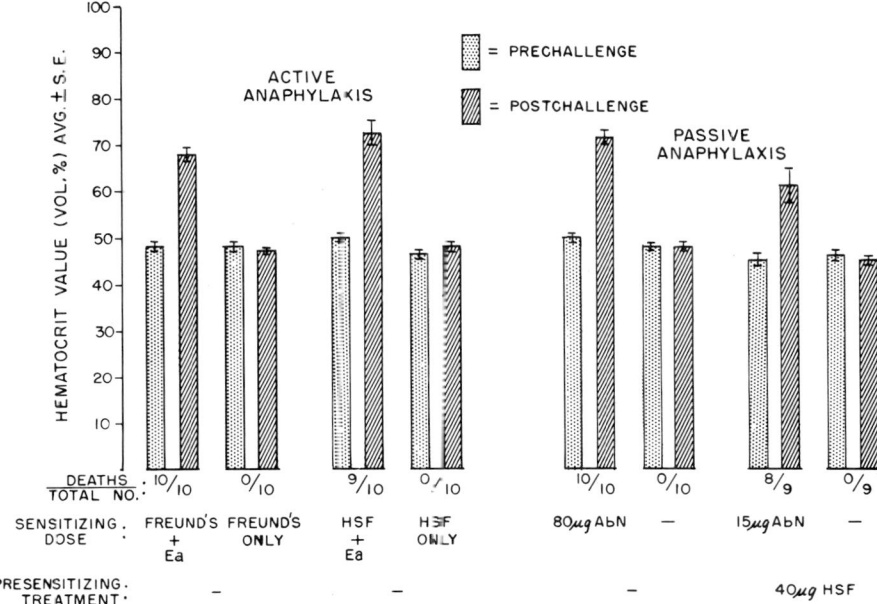

FIG. 9. Effect of active and passive anaphylaxis on hematocrit values in mice. (Reprinted from Ref. 97, p. 2, by courtesy of Williams and Wilkins Co., copyright 1965.)

doses of pertussigen (98). Referring to the opening paragraph of this chapter then, we can best characterize anaphylactic shock or histamine shock in the mouse as a combination of hypovolemic shock and low-resistance shock. Pertussigen-treated mice in shock apparently have a significant loss of fluid from their vascular system and are unable to compensate sufficiently to maintain blood pressure and venous return to the heart.

The agents which are released in mice undergoing anaphylactic shock have not been identified with certainty. The 2 most likely substances are histamine and serotonin (18,19,99) although other factors have been suspected (100). Serotonin is unusual in rodents, because it promotes capillary permeability more effectively than histamine (101). Fink (21) and Tokuda and Weiser (23) found evidence which suggested strongly that serotonin is a mediator of

TABLE 5

Average Number of Injections and Volume of Plasma Extender Given to Mice to Promote Survival from Anaphylaxis and Histamine Toxicity[a]

Experiment	Dextran solution			Physiological saline			Untreated, S/T
	Avg. no. injections/ mouse	Avg. vol/ mouse (ml)	S/T[b]	Avg. no. injections/ mouse	Avg. vol/ mouse (ml)	S/T	
Active anaphylaxis							
0.5 mg of HEA with Freund's incomplete adjuvant	2	0.82	9/9	2	0.81	9/10	2/10
0.5 mg of HEA with HSF	3	1.10	4/7	3	1.28	1/8	0/9
Passive anaphylaxis							
Normal mice, 80 μg of Ab N	1	0.57	9/10	2	0.81	9/9	2/10
HSF-treated mice, 15 μg of Ab N	3	1.27	6/9	4	1.79	7/9	1/10
HSF-treated mice challenged with 0.5 mg of histamine	3	1.08	8/13	3	1.23	9/10	1/10

[a]Data from Ref. 97.
[b]Survivors/total challenged.

III. HYPOTHESES ABOUT SHOCK ENHANCEMENT

anaphylaxis in the mouse. Although the responses to anaphylaxis, histamine, and serotonin resemble each other in the mouse, some observations indicate that pertussigen does not make mice more susceptible to anaphylaxis by increasing their susceptibility to these substances. The increased sensitivity to histamine and serotonin does not correspond to the development of either active or passive anaphylaxis (12). Doses of antihistamine or antiserotonin drugs which are protective against histamine and serotonin toxicity are not protective against anaphylactic shock; it requires much larger, almost toxic levels, to protect against anaphylaxis (23,102). In fact, antiserotonin drugs did not protect against anaphylaxis in *B. pertussis*-treated mice. In considering the various agents which may mediate anaphylactic shock in mice one should not overlook the possible involvement of the kinins (bradykinin), SRS (slow-reacting substance), and prostaglandins (PGS and PGE). These are powerful tissue hormones that are known to produce vascdilatation, capillary permeability, and inflammation, but very little is known of their normal physiological functions.

C. Adrenergic Involvement

A number of the more recent hypotheses proposed for the mechanism of action of pertussigen have implicated the adrenal hormones and the adrenergic nervous system. For example, it was reported several years ago that adrenalectomized mice are also susceptible to histamine, serotonin, anaphylaxis, endotoxin, cold stress and other noxae (39,99,103-108). Initially, *B. pertussis* was thought to act directly on the adrenal gland, but there was no evidence of histological lesions in the gland (109). A decrease in ascorbic acid content of the adrenal gland (110) and some histological evidence of stress (111) have been reported. Intraperitoneal administration of water-insoluble preparations of adrenal steroids such as hydrocortisone and cortisone protected mice from histamine shock. However, the doses required for protection were 1 to 4 mg per mouse given 16 to 24 h before challenge (112,113). These unphysiological doses were not 100% protective, and probably their protective effects were nonspecific, as is that afforded by reticuloendothelial system blocking agents against anaphylaxis (92) and serotonin shock (93). When it was reported that adrenal demedullated mice were as sensitive to histamine as were totally adrenalectomized mice (Fig. 10), serious doubts were cast upon the direct involvement of steroid hormones in this phenomenon. Blockade against the activity of adrenal catecholamines in pertussigen-treated mice (98,114) appeared likely. Whereas adrenalectomized or adrenal-demedullated mice were easily protected from histamine death by administering 1.25 to 5.00 µg of l-epinephrine, it required 5.00 to 7.50 µg of l-epinephrine to protect pertussigen-treated mice, and even these doses were not very effective if a large quantity of pertussigen had been used (98)(Fig. 11). For example, when histamine sensitivity was induced by a dose of 2.5 µg of saline extract from *B. pertussis,* 7.5 µg of l-epinephrine protected mice, but it was of no avail against sensitization with 20 µg of saline extract. Thus, it was clear that pertussigen was acting to block the effects of catecholamines on certain tissues, rather than interfering with the release of hormones from the adrenal glands.

The hypothesis that pertussigen acts by blocking the response of certain effector organs or tissues to adrenal catecholamines received considerable support when Fishel et al. (49,84,115) reported that treatment of mice with β-adrenergic blocking agents (DCI, pronethalol) also produced histamine sensitivity. Most of the known β-adrenergic blocking agents are structural analogs of the β-stimulator, isoproterenol, and they most likely react with the same adrenergic receptors. However, considerable differences exist among β-adrenergic blocking agents in their potency to induce histamine sensitivity (116). As shown in Table 6, propranolol was much more potent than 3 other agents.

In addition to the similar effects of pertussigen treatment and β-adrenergic blockade on histamine sensitivity in mice, Fishel et al. (49,84,115) reported that pertussigen treatment mimicked metabolic

III. HYPOTHESES ABOUT SHOCK ENHANCEMENT

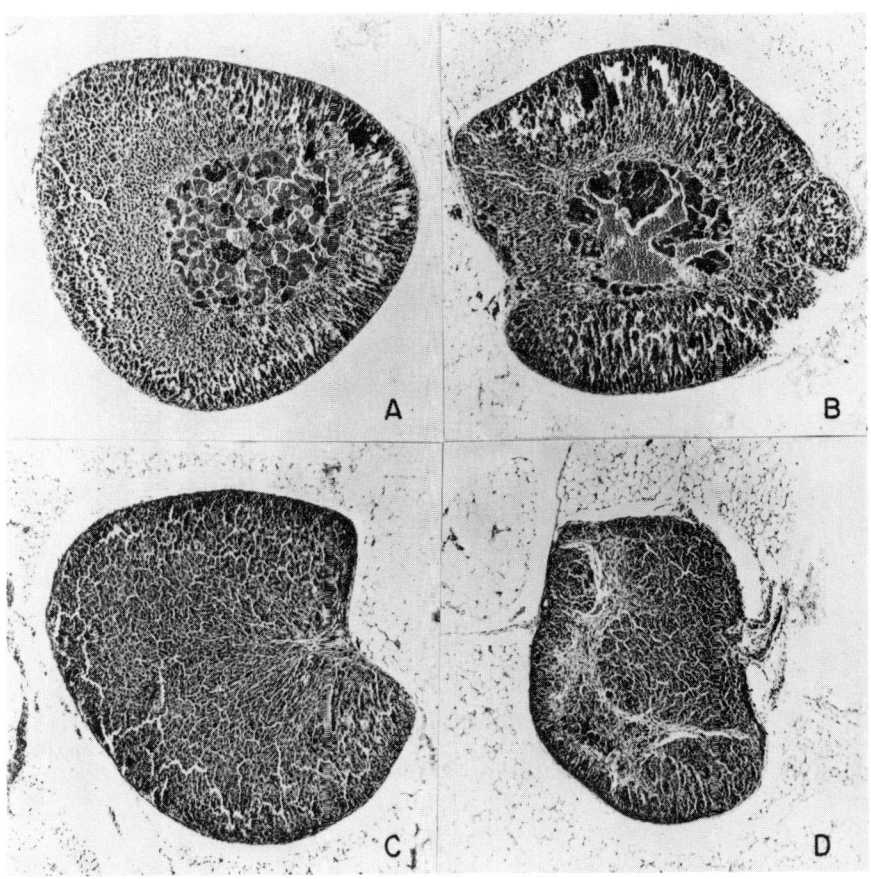

FIG. 10. Cross sections of adrenal glands from sham operated and demedullated mice. (A) Adrenal gland from a sham operated mouse which survived histamine challenge (note medullary tissue). (B) Demedullated adrenal gland from mouse which survived histamine challenge (note medullary tissue). (C and D) Demedullated adrenal glands from mice which died from histamine challenge (note absence of medullary tissue). (Reprinted from Ref. 98, p. 430, by courtesy of the Society for Experimental Biology and Medicine.)

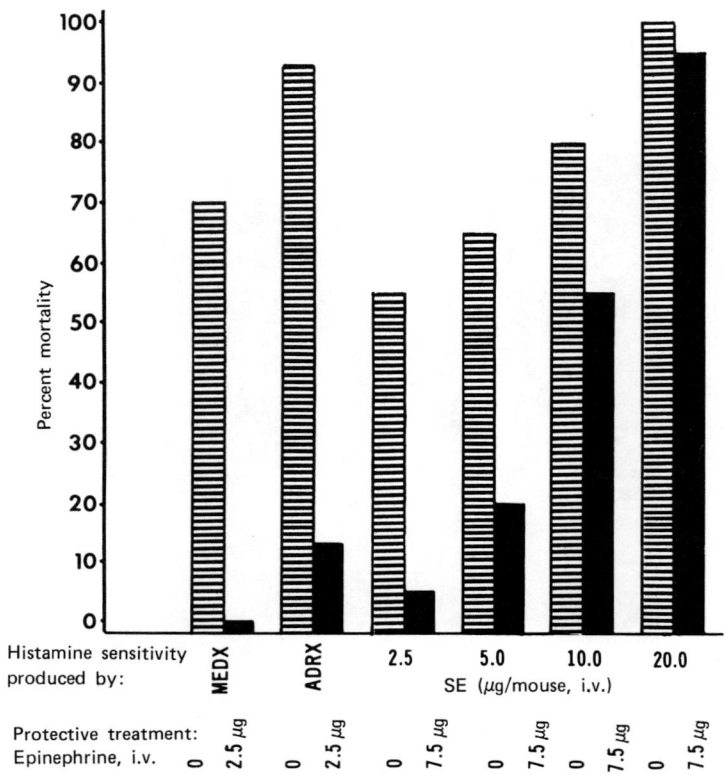

FIG. 11. Protection against histamine by epinephrine given within 30 sec after histamine challenge in demedullated (MEDX), adrenalectomized (ADRX), and mice given saline extract (SE) from *B. pertussis*. (Data from Ref. 98.)

changes produced in mice by the β-adrenergic blocking agents, especially on glucose metabolism. Both treatments produce hypoglycemia and block the ability of epinephrine to induce hyperglycemia. While β-adrenergic blocking agents administered to normal mice produced histamine sensitivity, the α-adrenergic blocking agent, dibenzyline, protected pertussigen-treated mice against histamine and serotonin (49). The foregoing observations led Fishel et al. to propose a hypothesis about the action of pertussigen. They suggested that the active material in *B. pertussis* cells acts directly to block β-adrenergic receptors or causes the animal to elaborate a material

III. HYPOTHESES ABOUT SHOCK ENHANCEMENT

TABLE 6

Comparison of the Efficacy of Various β-Adrenergic Blocking Agents in Inducing Histamine Sensitivity in Mice[a]

| Dose (µg/mouse i.v.) | DCI[b] | β-Adrenergic blocking agent | | Propranolol |
		Pronethalol	Butoxamine HCl	
240	7/10[c]	8/10	8/10	10/10
80	6/10	7/10	5/10	10/10
27	1/10	1/10	5/10	10/10
9	2/10	2/10	2/10	9/10
3	2/10	0/10	1/10	10/10
2	1/10	2/10	3/10	8/10
1	1/10	1/10	1/10	3/10
0.5	0/10	2/10	1/10	4/10
0.25	3/10	0/10	1/10	2/10
0 (Saline)		1/10		

[a]Reprinted from Ref. 116, p. 1174, by courtesy of Macmillan Journals Ltd.
[b]DCI = dichloroisoproterenol.
[c]Deaths/number of mice tested (all mice challenged i.p. with 0.5 mg histamine base, 20 min after receiving the blocking agent).

with a steric configuration complementary to the β-adrenergic receptors (115). The animal thus has a β-adrenergic blockade and when challenged with histamine or serotonin, the effects of endogenously released catecholamines are blocked at the β-receptor level leaving the agonistic effects on the α-receptors unopposed. This situation leads to an imbalanced response of the adrenergic receptors, with unfavorable metabolic responses plus smooth muscle and neural responses that decrease resistance to histamine and serotonin. This hypothesis has stimualted many studies into the role of the β-adrenergic nervous system and the effects of pertussigen on enzyme systems at the intracellular level. As originally proposed, the hypothesis is not supported by several subsequent observations.

We discovered a disparity between adrenergic agonists and their ability to protect against histamine sensitivity in β-adrenergic blocked mice and mice treated with pertussigen. Both epinephrine (an α- and β-receptor agonist) and isoproterenol (a β-receptor agonist) administered i.v. protect β-blocked mice against histamine challenge. Yet, only epinephrine protected pertussigen-treated mice while isoproterenol was deleterious and increased the mortality following histamine challenge (117)(Table 7). While there are certainly many similarities between the effects of pertussigen treatment and β-adrenergic blockade, especially in relation to metabolic effects, histamine sensitivity cannot be completely explained on the basis of β-adrenergic blockade.

Gulbenkian et al. (82) and Muszbek et al. (83) have observed that the hypoinsulinemia which occurs in normal rats following treatment with epinephrine is reversed in *B. pertussis*-treated rats and epinephrine produces hyperinsulinemia. Since increased insulin secretion is a β-activated response, the effect of epinephrine in pertussigen-treated rats is contrary to a postulated β-adrenergic blockade. In fact, this observation would suggest the possibility of an α-adrenergic blockade.

The involvement of the β-adrenergic nervous system and alterations of glucose metabolism have naturally led to investigations on the intracellular adenylate cyclase system and production of cyclic adenosine monophosphate (c-AMP). Results in this area have been somewhat confusing. Cronholm and Fishel (118) reported that c-AMP did not affect histamine sensitivity in normal CFW and CFI mice while 5'-AMP induced histamine hypersensitivity in CFW mice. Matsumura et al. (119) found that both of the above nucleotides administered to *B. pertussis*-treated female mice (HSL-SW [ICR]) protected against histamine. They produced optimum protection when given 90 min before the histamine challenge. We tried to repeat the findings of Matsumura et al. in our laboratory-reared strain of CFW mice and were unsuccessful (R. K. Bergman, unpublished observations). Apparently, as has been noted before, mouse strains differ distinctly in their susceptibility to histamine sensitization.

III. HYPOTHESES ABOUT SHOCK ENHANCEMENT

TABLE 7

Comparison of Protection Induced by Isoproterenol and Epinephrine to Histamine Challenge in Histamine-Sensitive Mice[a]

Catecholamine treatment	Histamine sensitivity induced by			
	Propranolol[b]		Saline extract[c]	
	D/T[d]	P[e]	D/T	P
None	17/20		10/20	
l-Epinephrine (μg/mouse)				
1.25	4/10	0.0184	--[f]	
2.50	4/10	0.0184	--	
5.00	0/10	<0.0049	1/20	0.0017
7.50	3/10	0.0049	0/20	0.0002
Isoproterenol (μg/mouse)				
0.3125	7/10	>0.025	--	
0.625	7/10	>0.025	--	
1.25	8/10	>0.025		
2.5	6/20	0.0005	9/20	0.5000
10.0	4/20	0.0000	14/20	0.1666
40.0	7/20	0.0015	15/20	0.0954
160.0	7/20	0.0015	19/20	0.0017
640.0	6/20	0.0005	20/20	0.0002

[a] Reprinted from Ref. 117, "Effects of Epinephrine, Norepinephrine and Isoproterenol Against Histamine Challenge in *Bordetella pertussis*-Treated and β-Adrenergic Blocked Mice," p. 564, by courtesy of Pergamon Press, Inc., copyright 1971.

[b] Mice were sensitized i.v. with 2 μg of propranolol and challenged i.p. with 0.5 mg of histamine 20 min later, followed by the i.v. dose of catecholamine 30 sec later.

[c] Mice were sensitized i.v. with 10 μg of saline extract of *B. pertussis* and challenged the next day with 0.25 mg of histamine, followed by the catecholamine dose 30 sec later.

[d] Deaths/total mice challenged.

[e] Control vs. catecholamine-treated mice. The levels of significance are: <0.025 = significant; <0.005 = highly significant.

[f] -- = not done.

Ortez et al. (120) found that mouse spleens from *B. pertussis*-treated mice had 44% less c-AMP than spleens from normal mice and that epinephrine induced 50% less c-AMP in sensitized mice than in normal mice. We tested mouse lymphocytes removed from lymph nodes of normal CFW mice and mice treated with pertussigen. Those from treated mice contained about 72% more c-AMP than did the lymphocytes from normal mice (R. K. Bergman, unpublished observations).

Parker and Morse (121) studied the c-AMP response to isoproterenol in suspensions of human blood leukocytes incubated with fractions of *B. pertussis* cells. Various fractions obtained from the cells reduced the production of c-AMP by the leukocytes after stimulation with isoproterenol. However, these same fractions were equally effective against prostaglandin E_1 and methacholine stimulation of c-AMP in leukocytes. Thus, the inhibition of c-AMP production in suspensions of human leukocytes is not strictly a β-adrenergic blockade phenomenon. These results led the authors to conclude: "Although these observations provide direct confirmation of the ability of *B. pertussis* to inhibit catecholamine responsiveness, the fact that PGE and methacholine responses were also inhibited suggests that blockade at the level of the β-adrenergic receptor is doubtful." (121).

That pertussigen increases the permeability of the vascular bed of certain tissues has been postulated. We (104) reported that if normal and *B. pertussis*-treated mice both received equal doses of Evans blue dye i.v., the treated mice had, 20 min to 4 h later, a lower serum concentration of dye than the controls. This observation suggested that the vascular bed of pertussigen-treated mice was more permeable to the dye. However, our discovery that pertussigen produces hypoalbuminemia (122) in mice cast some doubt on the permeability hypothesis. Evans blue dye attaches to serum albumin and forms a complex which remains in the circulation for a long time. If the quantity of dye administered exceeds the binding capacity of the serum albumin and the capillaries are freely permeable to unbound dye, the dye concentration in the serum would be directly related to the serum albumin concentration rather than to capillary permeability.

III. HYPOTHESES ABOUT SHOCK ENHANCEMENT

Nevertheless, we have felt for some time that the hypothesis of pertussigen acting by enhancing vascular permeability had some merit and if it could be proven it could possibly explain many of its actions. On various occasions we used different methods to investigate the effect of pertussigen on vascular permeability, but all were imprecise. Leibowitz and Kennedy (123) described a quantitative double isotope method for measuring vascular permeability that seems to resolve many of the difficulties found in other methods. Its greatest advantage is in that it is independent of changes in vascular blood volume. Using this method, we have determined that there is a significant increase in the permeability of the vascular bed in the thigh muscle of pertussigen-treated mice as compared with normal mice (R. K. Bergman, unpublished observations)(Table 8). The loss of ^{131}I-labeled human serum albumin or mouse serum albumin over a 24 h period was 59 to 71% greater in pertussigen-treated mice than in normal mice. Occasionally we have detected some increased permeability in the kidneys and lungs, but this has not been consistent and the results in the liver, spleen, and brain have always been negative. We have also investigated the effects of histamine and serotonin on vascular permeability when administered intracutaneously to normal and pertussigen-treated mice. Again, pertussigen significantly enhances the permeability inducing effects of these two amines (124). Interestingly, we found that when isoproterenol (β-adrenergic agonist) was administered i.v. immediately following histamine or serotonin given intracutaneously, permeability was markedly decreased. Epinephrine (α- and β-adrenergic agonist) and norepinephrine (α-adrenergic agonist) were less effective than isoproterenol by the i.v. route. When the catecholamines were administered concomitantly with histamine or serotonin in the skin test sites, then epinephrine was the most effective, norepinephrine less effective, and isoproterenol the least effective in blocking permeability. This observation again raises serious questions about the hypothesis of β-adrenergic blockade as being the means by which pertussigen acts to increase histamine and serotonin sensitivity.

TABLE 8

Effect of Pertussigen on Permeability of Vascular Bed in Mouse Thigh Muscle to Radiolabeled Human Serum Albumin and Radiolabeled Mouse Serum Albumin[a]

Permeability indicator	Treatment	EVBE[b] in thigh muscle over 24 h period (mean ± SEM)
^{131}I-HSA and ^{125}I-HSA	Saline controls	8.169 ± 0.188[c]
	20 µg SE of B. pertussis given i.v. e days before test	12.989 ± 0.417
^{131}I-MSA and ^{125}I-MSA	Saline controls	9.054 ± 0.526
	20 µg SE of B. pertussis given i.v. 3 days before test	15.478 ± 0.297

[a] Data from R. K. Bergman, unpublished observations.
[b] EVBE = extra vascular blood equivalents. For methodology see Appendix.
[c] Ten mice per treatment group.

D. Summary

At present, we might summarize our knowledge about the mechanism by which pertussigen enhances or promotes circulatory shock as follows: Pertussigen acts in some manner to block the effects of catecholamines at important effector organs or tissues in the mouse. (Whether or not this blockade is specifically directed to the β-adrenergic receptors is not certain.) Consequently, when the animal is challenged with histamine, serotonin, or some other noxae which produce vasodepression and increased vascular permeability, it is not capable, as normal mice can, to make the necessary compensations to sustain life and dies of shock. The fact that pertussigen itself apparently acts upon the vascular bed of the striated muscle mass of the mouse body to increase vascular permeability places an additional burden upon the homeostatic mechanisms of the circulatory system producing a serious diminution of vascular integrity.

REFERENCES

1. W. F. Ganong, *Review of Medical Physiology*, 6th ed., Lange Medical Publications, Los Altos, Calif., 1973, p. 465.
2. G. Eldering, *Amer. J. Hyg.*, *36*, 294 (1942).
3. A. G. Ospeck and M. E. Roberts, *J. Infect. Diseases*, 74, 22 (1944).
4. I. A. Parfentjev, M. A. Goodline, and M. E. Virion, *J. Bacteriol.*, *53*, 603 (1947).
5. I. A. Parfentjev, M. A. Goodline, and M. E. Virion, *J. Bacteriol.*, *53*, 613 (1947).
6. I. A. Parfentjev, M. A. Goodline, and M. E. Virion, *J. Bacteriol.*, *53*, 597 (1947).
7. S. Malkiel and B. J. Hargis, *Proc. Soc. Exptl. Biol. Med.*, *80*, 122 (1952).
8. S. Malkiel and B. J. Hargis, *Proc. Soc. Exptl. Biol. Med.*, *81*, 109 (1952).
9. S. Malkiel and B. J. Hargis, *J. Allergy*, *23*, 352 (1952).
10. S. Malkiel and B. J. Hargis, *Proc. Soc. Exptl. Biol. Med.*, *81*, 689 (1952).
11. J. Munoz and R. K. Bergman, *Bacteriol. Rev.*, *32*, 103 (1968).
12. J. Munoz and R. L. Anacker, *J. Immunol.*, *83*, 502 (1959).
13. J. Munoz, L. F. Schuchardt, and W. F. Verwey, *J. Immunol.*, *80*, 77 (1958).
14. M. Pittman and F. G. Germuth, *Proc. Soc. Exptl. Biol. Med.*, *87*, 425 (1954).
15. J. Munoz, L. F. Schuchardt, and W. F. Verwey, *Fed. Proc.*, *13*, 507 (1954).
16. I. A. Parfentjev and M. A. Goodline, *J. Pharmacol. Exptl. Therap.*, *92*, 411 (1948).
17. J. Munoz, *Proc. Soc. Exptl. Biol. Med.*, *95*, 328 (1957).
18. M. Pittman, *Fed. Proc.*, *16*, 867 (1957).
19. L. S. Kind, *Proc. Soc. Exptl. Biol. Med.*, *95*, 200 (1957).

20. P. Kallos and L. Kallos-Deffner, *Arch. Allergy Appl. Immunol.*, *11*, 237 (1957).
21. M. A. Fink, *Proc. Soc. Exptl. Biol. Med.*, *92*, 673 (1956).
22. M. D. Gershon and L. L. Ross, *J. Exptl. Med.*, *115*, 367 (1962).
23. S. Tokuda and R. S. Weiser, *J. Immunol.*, *86*, 292 (1961).
24. R. N. Arch and I. A. Parfentjev, *J. Infect. Diseases*, *101*, 31 (1957).
25. R. J. Dubos and R. W. Schaedler, *J. Exptl. Med.*, *104*, 53 (1956).
26. G. Eldering, *Amer. J. Hyg.*, *36*, 294 (1942).
27. I. A. Parfentjev, *6th Intern. Congr. Microbiol., Rome*, *3*, 38 (1953).
28. I. A. Parfentjev, *Proc. Soc. Exptl. Biol. Med.*, *90*, 373 (1955).
29. R. S. Abernathy and W. W. Spink, *J. Immunol.*, *77*, 418 (1956).
30. L. Chedid, *Ann. Endocrinol. (Paris)*, *15*, 746 (1954).
31. L. Chedid and F. Boyer, *Ann. Inst. Pasteur*, *94*, 341 (1958).
32. L. S. Kind, *J. Immunol.*, *82*, 32 (1959).
33. J. G. Michael, *Proc. Soc. Exptl. Biol. Med.*, *113*, 495 (1963).
34. I. A. Parfentjev, *Yale J. Biol. Med.*, *27*, 46 (1954).
35. R. E. Pieroni and L. Levine, *J. Allergy*, *39*, 93 (1967).
36. L. Levine, *Symp. Series Immunobiol. Standard.*, *3*, 129 (1967).
37. M. Rowen, W. S. Moos, and M. Samter, *Proc. Soc. Exptl. Biol. Med.*, *88*, 548 (1955).
38. L. S. Kind and R. H. Gadsden, *Proc. Soc. Exptl. Biol. Med.*, *84*, 373 (1953).
39. J. Munoz and L. F. Schuchardt, *Proc. Soc. Exptl. Biol. Med.*, *94*, 186 (1957).
40. G. Hunder and W. W. Spink, *Proc. Soc. Exptl. Biol. Med.*, *95*, 55 (1957).
41. S. Malkiel and B. J. Hargis, *J. Allergy*, *31*, 508 (1960).
42. R. E. Pieroni and L. Levine, *J. Allergy*, *39*, 25 (1967).
43. S. Malkiel and B. J. Hargis, *Proc. Soc. Exptl. Biol. Med.*, *125*, 565 (1967).
44. R. G. Townley, I. L. Trapani, and A. Szentivanyi, *J. Allergy*, *39*, 177 (1967).
45. A. Guerault and M. Quevillon, *Bacteriol. Proc.*, p. 52, (1965).
46. H. B. Maitland, R. Kohn, and A. D. MacDonald, *J. Hyg.*, *53*, 196 (1955).
47. M. G. Stronk and M. Pittman, *J. Infect. Diseases*, *96*, 152 (1955).

REFERENCES

48. R. K. Bergman and J. Munoz, *Int. Arch. Allergy*, *34*, 331 (1968).
49. C. W. Fishel, A. Szentivanyi, and D. W. Talmage, *J. Immunol.*, *89*, 8 (1962).
50. I. A. Parfentjev, *Proc. Amer. Assoc. Cancer Res.*, *2*, 38 (1955).
51. R. E. Pieroni, E. J. Broderick, and L. Levine, *J. Immunol.*, *95*, 643 (1965).
52. J. Munoz and L. F. Schuchardt, *J. Allergy*, *24*, 330 (1953).
53. L. S. Kind, *J. Immunol.*, *77*, 118 (1956).
54. S. Malkiel and B. J. Hargis, *J. Allergy*, *29*, 524 (1958).
55. R. K. Bergman and J. Munoz, *Proc. Soc. Exptl. Biol. Med.*, *117*, 400 (1964).
56. R. K. Bergman and J. Munoz, *Int. Arch. Allergy*, *34*, 9 (1968).
57. G. F. Gauthier, E. R. Loew, and H. J. Jenkins, *Proc. Soc. Exptl. Biol. Med.*, *90*, 726 (1955).
58. I. A. Parfentjev, *Proc. Soc. Exptl. Biol. Med.*, *89*, 297 (1955).
59. L. S. Kind, *J. Immunol.*, *70*, 411 (1953).
60. S. Malkiel and B. J. Hargis, *J. Allergy*, *31*, 513 (1960).
61. A. Guerault, *Round Table Conference on Pertussis Immunization, Prague*, *2*, 299 (1962).
62. M. Pittman, *J. Infect. Diseases*, *89*, 296 (1951).
63. J. Munoz, in *Bacterial Endotoxins* (M. Landy and W. Braun, eds.), Rutgers University Press, New Brunswick, N.J., 1954, p. 460.
64. M. Pittman, *J. Infect. Diseases*, *89*, 300 (1951).
65. J. Munoz, L. F. Schuchardt, and W. F. Verwey, *J. Allergy*, *25*, 120 (1954).
66. S. Malkiel, B. J. Hargis, and S. M. Feinberg, *J. Immunol.*, *71*, 311 (1953).
67. J. Munoz and R. K. Bergman, *J. Immunol.*, *97*, 120 (1966).
68. I. Joó, Z. Pusztai, and V. P. Juhász, *Z. Immunitaetsforsch.*, *121*, 159 (1961).
69. M. Pittman, *Proc. Soc. Exptl. Biol. Med.*, *77*, 70 (1951).
70. B. D. Geller and M. Pittman, *Infect. Immun.*, *8*, 83 (1973).
71. L. S. Kind and E. F. Woods, *Proc. Soc. Exptl. Biol. Med.*, *84*, 601 (1954).
72. T. Matsui, M. Kishigami, and Y. Kuwajima, *Nature*, *183*, 756 (1959).
73. T. Matsui, M. Kishigami, and Y. Kuwajima, *J. Hyg. Epidemiol. Microbiol. Immunol. (Prague)*, *4*, 108 (1960).
74. M. Niwa, M. Nakamura, and Y. Kuwajima, *Compt. Rend. Soc. Biol.*, *159*, 2098 (1965).

75. M. Niwa, Y. Yamadeya, K. Hamada, and Y. Kuwajima, *Jap. J. Bacteriol.*, *14*, 1026 (1959).
76. M. Niwa, Y. Yamadeya, T. Matsui, and Y. Kuwajima, *Nature*, *183*, 755 (1959).
77. M. Niwa, Y. Yamadeya, and Y. Kuwajima, *J. Infect. Diseases*, *112*, 107 (1963).
78. R. W. Schayer, in *Histamine* (G. E. W. Wolstenholme and C. M. O'Conner, eds.), Ciba Foundation Symposium, Little, Brown and Co., Boston, p. 183.
79. R. W. Schayer and O. H. Ganley, *J. Allergy*, *31*, 204 (1961).
80. R. W. Schayer and O. H. Ganley, *Amer. J. Physiol.*, *197*, 721 (1959).
81. A. Szentivanyi, S. Katsh, and B. McGarry, *Fed. Proc.*, *27*, p. 268 (1968).
82. A. Gulbenkian, L. Schobert, C. Nixon, and I. I. A. Tabachnick, *Endocrinology*, *83*, 885 (1968).
83. L. Muszbek, B. Csaba, and J. Csongor, *Acta Allergologica*, *28*, 138 (1973).
84. C. W. Fishel and A. Szentivanyi, *J. Allergy*, *34*, 439 (1963).
85. A. Gulbenkian, A. Y. Grasso, and I. A. Tabachnick, *Biochem. Pharmacol.*, *16*, 783 (1967).
86. I. A. Parfentjev and W. L. Schleyer, *Arch. Biochem.*, *20*, 341 (1949).
87. A. Szentivanyi, C. W. Fishel, and D. W. Talmage, *J. Infect. Diseases*, *113*, 86 (1963).
88. K. F. Keller and C. W. Fishel, *J. Bacteriol.*, *94*, 804 (1967).
89. B. Gözsy and L. Kátó, *Rev. Can. Biol.*, *2*, 427 (1964).
90. R. K. Bergman and J. Munoz, *Proc. Soc. Exptl. Biol. Med.*, *131*, 42 (1969).
91. L. S. Cronholm and C. W. Fishel, *J. Bacteriol.*, *95*, 1993 (1968).
92. R. Wistar, P. E. Treadwell, and A. F. Rasmussen, *J. Exptl. Med.*, *111*, 631 (1960).
93. O. H. Ganley, *Bact. Proc.*, p. 88 (1960).
94. P. D. McMaster and H. Kruse, *J. Exptl. Med.*, *89*, 583 (1949).
95. J. D. Fulton, W. E. Harris, and C. E. Craft, *Proc. Soc. Exptl. Biol. Med.*, *95*, 625 (1957).
96. J. Munoz and R. K. Bergman, *Nature*, *205*, 199 (1965).
97. R. K. Bergman and J. Munoz, *J. Immunol.*, *95*, 1 (1965).
98. R. K. Bergman and J. Munoz, *Proc. Soc. Exptl. Biol. Med.*, *122*, 428 (1966).

REFERENCES

99. B. N. Halpern, *Compt. Rend. Soc. Biol.*, *146*, 1996 (1952).
100. A. Olivera Lima, *Int. Arch Allergy*, *32*, 46 (1967).
101. W. H. Douglas, in *The Pharmacological Basis of Therapeutics* (L. S. Goodman and A. Gilman, eds.), Macmillan Co., New York, 1965, p. 644.
102. J. Munoz and L. F. Schuchardt, *Fed. Proc.*, *14*, 473 (1955).
103. L. Chedid, *Ann. Endocrinol. (Paris)*, *15*, 746 (1954).
104. J. Munoz, *J. Immunol.*, *86*, 618 (1961).
105. J. Munoz and L. F. Schuchardt, *J. Allergy*, *25*, 125 (1954).
106. D. Perla and J. Marmorston, in *Natural Resistance and Clinical Medicine*, Little Brown and Co., Boston, 1941.
107. G. Pincus and K. V. Thimann, in *The Hormones: Physiology, Chemistry and Applications*, Academic Press, Inc., New York, 1948.
108. J. Munoz, L. F. Schuchardt, and W. F. Verwey, *Fed. Proc.*, *13*, 507 (1954).
109. S. Malkiel, *J. Allergy*, *27*, 445 (1956).
110. J. Pekárek and K. Řežabek, *J. Hyg. Epidemiol. Microbiol. Immunol.*, *3*, 79 (1959).
111. M. Nakamura, *J. Osaka City Med. Center*, *11*, 295 (1962).
112. L. Chedid and F. Boyer, *Ann. Inst. Pasteur*, *91*, 380 (1956).
113. L. S. Kind, *J. Allergy*, *24*, 52 (1953).
114. R. K. Bergman and J. Munoz, *Nature*, *205*, 910 (1965).
115. C. W. Fishel, A. Szentivanyi, and D. W. Talmage, in *Bacterial Endotoxins*, (M. Landy and W. Braun, eds.), Rutgers University Press, New Brunswick, N.J., 1964, p. 474.
116. R. K. Bergman and J. Munoz, *Nature*, *217*, 1173 (1968).
117. R. K. Bergman and J. Munoz, *Life Sci.*, *10*, 561 (1971).
118. L. S. Cronholm and C. W. Fishel, *Proc. Soc. Exptl. Biol. Med.*, *127*, 1178 (1968).
119. Y. Matsumura, J. H. Vaughan and E. M. Tan, *J. Allergy Clin. Immunol.*, *54*, 191 (1974).
120. R. A. Ortez, T. W. Klein, and A. Szentivanyi, *J. Allergy*, *45*, 111 (1970).
121. C. W. Parker and S. I. Morse, *J. Exptl. Med.*, *137*, 1078 (1973).
122. R. K. Bergman and J. Munoz, *Proc. Soc. Exptl. Biol. Med.*, *131*, 964 (1969).
123. S. Leibowitz and L. Kennedy, *Immunology*, *22*, 859 (1972).
124. R. K. Bergman and J. Munoz, *J. Allergy Clin. Immunol.*, *55*, 378 (1975).

Chapter 4

ENHANCEMENT OF ANAPHYLACTIC SENSITIVITY

I.	GENERAL REMARKS	109
II.	ACTIVE ANAPHYLAXIS	110
III.	PASSIVE ANAPHYLAXIS	114
IV.	OTHER HYPERSENSITIVITY REACTIONS	118
	A. Passive Cutaneous Anaphylaxis	118
	B. Schultz-Dale Reaction	118
V.	MECHANISM OF PERTUSSIGEN ACTION IN ANAPHYLAXIS	119
	REFERENCES	121

I. GENERAL REMARKS

In the previous chapter the shock-enhancing effect of pertussigen was discussed in relation to its ability to decrease the resistance or tolerance of mice to chemical mediators of shock. Anaphylaxis is a unique form of shock produced by an interaction between antibody and antigen in the animal body, resulting in the endogenous release of shock-inducing mediators. For this reason the enhancing effect of pertussigen on this form of shock is discussed separately. Furthermore, in this chapter the effect of pertussigen on immediate hypersensitivity reactions unrelated to generalized shock is also discussed.

II. ACTIVE ANAPHYLAXIS

Compared to guinea pigs, mice have a lower and more irregular susceptibility to anaphylactic shock (1). This resistance can be overcome by intense immunization or by the use of adjuvants (2-6) of which the most effective is pertussis vaccine. As noted in the previous chapter, the first to observe this phenomenon were Eldering (7), Ospeck and Roberts (8), and Parfentjev et al. (9). However, this sensitivity was not clearly recognized as anaphylactic shock until Malkiel and Hargis (10,11) showed that antigens administered with pertussis vaccine induced a marked sensitivity to fatal anaphylaxis. These authors induced anaphylactic sensitivity in mice by giving i.p. 0.03 ml horse serum mixed with 8×10^9 B. pertussis killed cells. Upon challenge 15 days later with 0.1 ml of horse serum, 95 of 102 mice died of shock, compared with only 3 of 66 control mice sensitized with horse serum alone. Similar results were obtained with other antigens. Our results show that crude alkaline saline extracts from acetone-dried B. pertussis cells are as effective as whole B. pertussis cells in inducing anaphylactic sensitivity in mice (Table 1). As little as 1 µg of semipurified pertussigen mixed with 500 µg of HEA induced marked anaphylactic sensitivity (Table 2). The enhancing factor is inactivated by heating at 80°C for ½ h (Fig. 1) and thus is different from endotoxin which is also somewhat effective in increasing anaphylactic susceptibility in mice (13).

Anaphylactic sensitivity is most effectively induced by giving pertussigen mixed with antigen i.p. or i.v., but sensitization can also be induced when pertussigen is given by a route different from the antigen (Table 3). Many soluble antigens have been used to study anaphylaxis in the pertussis vaccine-treated mouse. We have used HEA and bovine serum albumin (BSA), while others have used horse serum, hemocyanin, bovine or human gamma globulin, and lysozyme. Some workers have used particulate antigens such as chicken erythrocytes (14).

TABLE 1

Effect of B. pertussis Alkaline Saline Extract (SE) on Anaphylaxis in CFW Strain of Mice[a]

Sensitizing dose	Anaphylaxis (D/T)[b]
1 mg HEA + 40 µg SE	32/32
1 mg HEA + saline	0/20
Saline + 40 µg SE	0/19
Saline	0/20

[a] Mice were sensitized i.p. with 0.5 ml of saline containing the indicated amounts of HEA and SE. They were challenged i.v. 22 days after sensitization with 0.5 mg of HEA in 0.2 ml (J. J. Munoz and R. K. Bergman, unpublished observations).

[b] Deaths/total mice challenged.

TABLE 2

Effectiveness of Pertussigen to Induce Active Anaphylaxis to HEA in CFW Mice[a,b]

Sensitizing dose		Anaphylaxis (D/T)[d]
Pertussigen[c] (µg)	HEA (µg)	
5	500	10/10
1	500	6/10
0.2	500	3/10
0.04	500	0/10
5	--	0/10
--	500	0/10

[a] Data from J. J. Munoz and R. K. Bergman, unpublished observations.

[b] Mice (5 male and 5 female) received i.p. the indicated dose in 0.2 ml challenged i.v. 14 days later with 500 µg HEA.

[c] Prepared by method described in Ref. 12.

[d] Deaths/total mice tested.

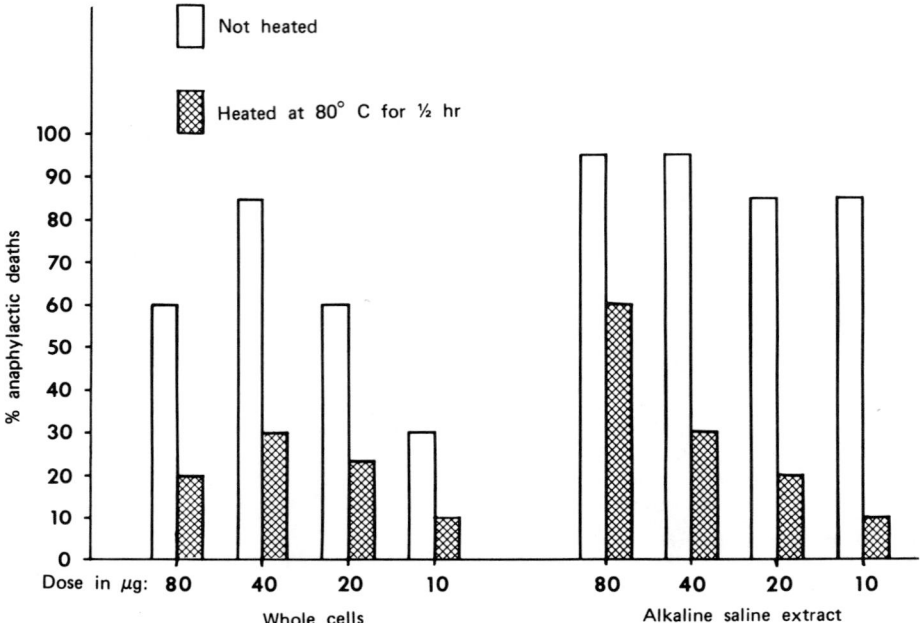

FIG. 1. Effect of heating whole cells or alkaline saline extracts of *B. pertussis* on their ability to promote anaphylactic sensitivity to HEA.

Swiss Webster albino mice, which have been used for most of our studies, are usually uniformly sensitized to anaphylaxis. However, some albino strains are irregularly susceptible to anaphylaxis, even when pertussigen is used. The ability of pertussigen to promote anaphylaxis is as great as that of Freund's complete adjuvant, although with Freund's complete adjuvant less antigen is capable of sensitizing mice. Pertussigen, however, induces a more uniform sensitization and once sensitivity is established, on or about the 7th day after a single sensitizing dose, it remains high for over 100 days. Figure 2 compares results obtained with these 2 adjuvants. The anaphylactic sensitivity (judged entirely by death) decreases with time in animals receiving the antigen in complete Freund's adjuvant. It should be stressed that the antibody titers in these animals were 10- to 100-fold higher than in those receiving pertussis

II. ACTIVE ANAPHYLAXIS

TABLE 3

Effect of Pertussigen on Anaphylaxis to HEA
When Given by Different Routes[a,b]

500 µg HEA	25 µg pertussigen[c]	Challenge 15 days later (D/T)[d]
i.p.	i.v.	13/15
i.v.	i.p.	13/15
s.c.	i.v.	15/15
s.c.	i.p.	12/15
s.c.	s.c.	11/15
i.p.	i.p.	15/15
i.v.	i.v.	14/15
i.v.	--	0/15
i.p.	--	0/15
s.c.	--	1/15
--	--	0/10

[a] Data from J. J. Munoz and R. K. Bergman, unpublished observations.
[b] Mice received the dose indicated in 0.2 ml and 15 days later were challenged with 500 µg HEA given i.v. in 0.2 ml.
[c] Prepared by method described in Ref. 12.
[d] Deaths/total mice tested.

vaccine. It is not known if the high anaphylactic sensitivity induced by pertussigen is due to stimulation of a highly effective class of immunoglobulin or to physiological changes produced by pertussigen. It is known, as discussed later, that pertussigen stimulates production of antibodies in the IgE class, but the role of this antibody class in mouse anaphylaxis has not been well studied.

Steroids, antihistamines, antiserotonins, and tranquilizers protect against anaphylaxis in mice which have been sensitized to an antigen without the use of pertussigen (16). However, these drugs are not very effective in protecting against anaphylactic shock in pertussigen-treated mice.

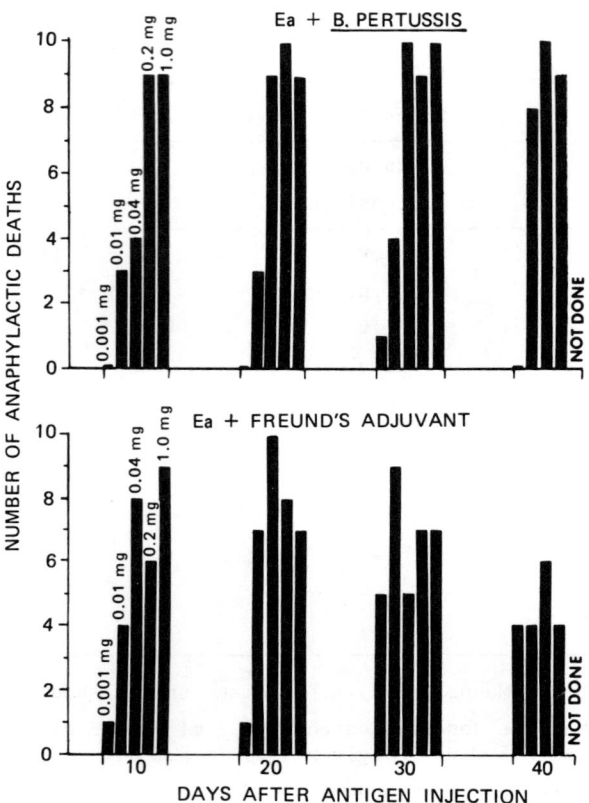

FIG. 2. Anaphylactic deaths in mice receiving various amounts of HEA mixed with with B. pertussis cells (top) or with Freund's adjuvant (bottom). Ten mice were used per group. The amount of sensitizing antigen is indicated in milligrams on top of the first set of bars at day 10; the same amounts were given for the other challenge days. Each mouse was challenged i.v. with 0.5 mg HEA in 0.2 ml of saline. (Reprinted from Ref. 15, p. 136, by courtesy of Williams and Wilkins Co., copyright 1963.)

III. PASSIVE ANAPHYLAXIS

Fatal anaphylaxis induced passively is, as in the case of active anaphylaxis, difficult to demonstrate in the mouse. Many workers have had success in inducing passive sensitization with heterologous antibodies, but others have not. Among the mice, in which passive

III. PASSIVE ANAPHYLAXIS

anaphylaxis was successfully produced with heterologous antibody, there was only a low percentage of fatal shock (17,18). However, when homologous (mouse) antibodies were used, fatal anaphylaxis was induced in a large percentage of mice (19). Administration of pertussis vaccine induces a greater sensitivity to passive anaphylaxis. Pittman and Germuth (20) and we (21,22) independently observed this phenomenon. Our observations with both heterologous and homologous antibody are illustrated in Table 4. Administration of 2×10^9 B. pertussis cells 4 days before injection of anti-BSA rabbit serum induced a marked sensitivity to the BSA challenge given 6 h after the antiserum. The SD_{50} of anti-BSA was more than 435 µg antibody nitrogen in normal mice, while in B. pertussis-treated mice it was only 160 µg. When mouse anti-HEA was used, the SD_{50} for normal mice was

TABLE 4

Passive Anaphylaxis Induced in B. pertussis-Treated Mice With Rabbit and Mouse Antibody[a]

Rabbit antibody			Mouse antibody		
Anti-BSA ab N/mouse (µg)	Normal	B. pertussis-treated[b]	Anti-HEA ab N/mouse (µg)	Normal	B. pertussis-treated
27	0/10	0/10	1.2	0/10	0/10
54	0/10	1/10	2.3	0/10	0/10
108	2/10	3/10	4.7	1/10	4/10
187	0/10	7/10	9.4	2/10	7/10
217	4/10	9/10	18.8	2/10	10/10
435	4/10	9/10	37.5	6/10	10/10
SD_{50}	<435 µg	160 µg	SD_{50}	26 µg	6.2 µg

[a]Reprinted from Ref. 19, p. 503, by courtesy of Williams and Wilkins Co., copyright 1959.
[b]Mice treated with B. pertussis received 2×10^9 cells i.p. 4 days before the i.p. injection of the desired amount of sensitizing antibody. Normal mice received only the sensitizing dose of antibody. Five to 6 h later the animals were challenged i.v. with 500 µg HEA or 1,000 µg BSA.

26 μg antibody nitrogen and in *B. pertussis*-treated mice it was only 6.2 μg. Anaphylaxis induced by administration of exogenously preformed antibody-antigen complexes was also enhanced by *B. pertussis* vaccine (23).

Large doses of pertussigen are not required to enhance passive anaphylaxis. As little as 250 million *B. pertussis* cells of some vaccines induce increased sensitivity and as little as 5 μg of semipurified pertussigen per mouse is effective.

The type of antibody involved in passive anaphylaxis is not exactly known, but in our work with mouse anti-HEA (24) the sera used did not contain IgE with specificity to the antigen because passive cutaneous anaphylaxis (PCA) reactions could not be elicited 24 h after sensitization. By immunoelectrophoresis, as seen in Fig. 3, the only demonstrable antibodies were in the IgG_1 and IgG_2 region. [In the mouse these 2 immunoglobulins have very similar electrophoretic mobilities (25).] When semipurified IgE preparations containing 1,000 PCA units (1 PCA = smallest amount of antibody that will produce a 5 mm diameter PCA reaction in the skin) were administered i.v. to mice, no fatal anaphylaxis occurred when the mice were challenged 6 h later with 0.5 mg HEA. However, a similar test utilizing 1,000 PCA units of IgG_1 antibody did produce fatal anaphylaxis (J. J. Munoz and R. K. Bergman, unpublished observations).

The effects of pertussigen on active anaphylaxis differ significantly from its effects on passive anaphylaxis. As pointed out before, once actively induced sensitivity to anaphylaxis is established in the pertussigen-treated mouse, the sensitivity remains high for over 100 days. Increased susceptibility to passive anaphylaxis after pertussigen administration can be shown only for about 1 week (22). In contrast, histamine sensitivity lasts for a much longer period of time (26). Therefore, there seems to be no direct relationship between increased sensitivity to histamine and increased susceptibility to passively induced anaphylaxis. After administration of rabbit antibody to pertussigen-treated mice, fatal sensitivity can be detected for only 48 h, while that induced with mouse antibody can be demonstrated 6 days later, indicating a more rapid metabolism of rabbit antibody (19).

III. PASSIVE ANAPHYLAXIS 117

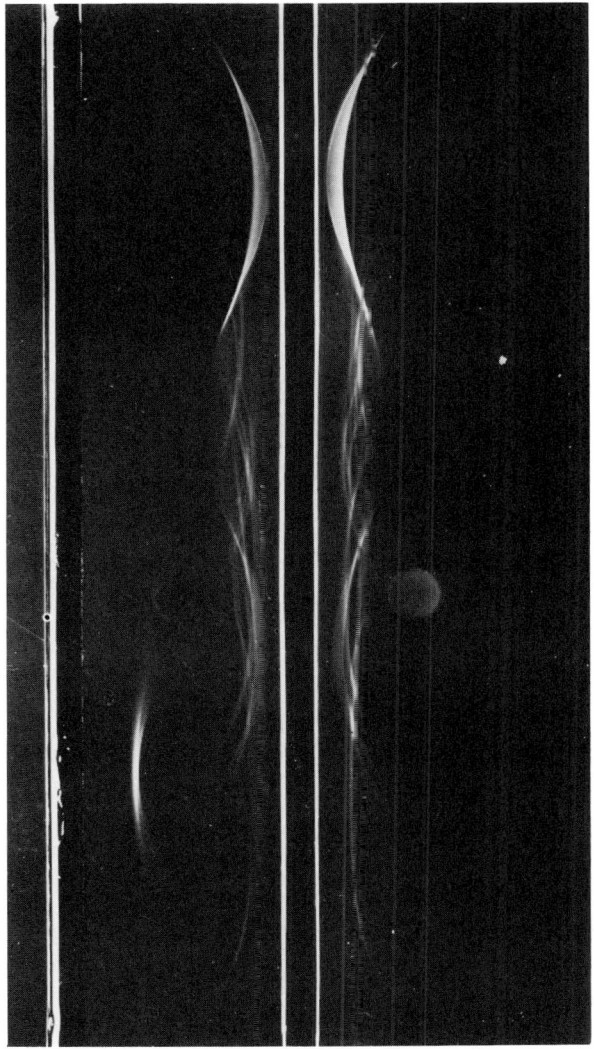

FIG. 3. Immunoelectrophoresis test employing mouse peritoneal fluid (top well) and mouse serum (bottom well). Developed with hyperimmune rabbit anti-mouse serum (center trough) and HEA (top trough). (Reprinted from Ref. 25, p. 429, by courtesy of Williams and Wilkins Co., copyright 1961.)

As mentioned in the previous chapter, pertussigen-treated mice react in many ways as adrenalectomized mice; this is also true in passive anaphylaxis. Adrenalectomized mice become highly susceptible to this type of shock (21,22). It is also important to note that strains of mice that do not become hypersensitive to histamine after *B. pertussis* (such as the CF-1 mouse) do not become sensitive to histamine after adrenalectomy (27). However, these mice do become sensitive to anaphylaxis (22), further indicating that the state of anaphylactic sensitivity does not depend on an increased susceptibility to histamine. It may, however, depend on an increased susceptibility to serotonin or to a combination of histamine and serotonin, or to other vasoactive agents.

IV. OTHER HYPERSENSITIVITY REACTIONS

A. Passive Cutaneous Anaphylaxis

The sensitivity of mice to PCA induced with antibody of the IgG_1 class (2 h PCA antibody) is not increased in mice by previous treatment with *B. pertussis*; in fact, we found that slightly more antibody is needed to demonstrate the PCA reactions in the skin of mice pretreated with *B. pertussis* extracts (24). This has not been studied with antibodies of the IgE type.

B. Schultz-Dale Reaction

The Schultz-Dale reaction in mice passively sensitized with mouse antibody to HEA was not modified by pretreatment of the mice with pertussis vaccine. However, as shown in Table 5, sensitivity of the uteri of mice to the antigen appeared sooner in animals receiving the antigen together with pertussis vaccine, which is a reflection of the adjuvant effect of this vaccine (28).

TABLE 5

Development of In Vitro Anaphylactic Sensitivity in Normal and
B. pertussis-Treated Mice Actively Sensitized to HEA[a]

Time after sensitization (days)	Results[b]		
	Normal sensitized	B. pertussis-treated	Normal unsensitized
5	0/5[c]	0/10	0/4
7	0/5	3/10	0/4
8	1/5	9/10	0/4
10	5/5	10/10	0/4

[a]Reprinted from Ref. 28, p. 72, by courtesy of the Society for Experimental Biology and Medicine.
[b]All muscle strips that gave a negative Schultz-Dale response did respond to 10 µg of acetycholine per ml of bath solution.
[c]Numerator = number of muscles with positive Schultz-Dale; denominator = number of muscles tested.

V. MECHANISM OF PERTUSSIGEN ACTION IN ANAPHYLAXIS

As described in Chapter 3, histamine and anaphylactic shock in the mouse are similar and due to circulatory failure as a result of loss of effective blood volume (29,30). This loss of effective blood volume must be due not only to dilatation of blood vessels but also to an increased vascular permeability which allows fluids to escape from the capillaries (31). This explains the edema and hemoconcentration observed. Some symptoms, such as the marked cyanosis and respiratory distress, must result from lack of oxygen in the tissues due to lack of blood circulation and not to constricted bronchioles in the lungs. Smooth muscle contraction does occur as indicated by the ruffling of hair, and urination. This picture of anaphylaxis can be observed in mice not treated with pertussis as well as in those receiving it. As discussed in Chapter 3, pertussigen treatment makes it more difficult for a mouse to recover from shock apparently due to an inability of epinephrine to act. This blockage of epinephrine action can explain the circulatory disorders which occur in passive anaphylaxis and, at least during the first few weeks after sensitization, in active anaphylaxis. In actively induced anaphylaxis other factors such as the adjuvant effect of *B. pertussis* must also enter into the picture; once sensitization is established the animal remains sensitive for a long time, without regard to its increased susceptibility to histamine. One intriguing observation is that increased sensitivity to passive anaphylaxis does not last long after pertussigen administration. The histamine sensitization effect of pertussigen is known to last much longer.

In pertussigen-treated mice sensitized to HEA, antihistamine and antiserotonin drugs do have some protective effect against anaphylaxis but this protection is not consistent and it is difficult to interpret. Antihistamines, however, are very effective in protecting against histamine shock in *B. pertussis*-treated mice. Likewise, antiserotonins protect against serotonin. In the Schultz-Dale reaction, the antiserotonin, lysergic acid diethylamide (LSD), prevents muscle contraction induced by contact with antigen (32), but we found it did not prevent fatal anaphylaxis in *B. pertussis*-treated mice (unpublished observations). It is interesting that very large doses of LSD (1,000 µg/mouse) when given i.p. 30 min before challenge completely inhibited the PCA reaction (unpublished observations). Large doses of the antihistamine, neoantergan, were not effective in inhibiting the PCA (unpublished observations). In anaphylaxis induced by soluble antigen-antibody complexes some of the antiserotonin drugs such as reserpine are effective (33). We found that reserpine also protected mice from anaphylaxis induced by passive means. A dose of 62 µg of reserpine given i.p. to mice 36 to 39 h prior to sensitization and challenge, protected mice from anaphylactic death. Thirty-six hours after reserpine treatment the rectal temperature of these mice, however, was much lower than that of untreated controls. The lowering of body temperature by reserpine may have pronounced effects on anaphylactic reactions since we have also shown that normal or *B. pertussis*-treated mice passively sensitized with mouse antibody are protected from anaphylactic death by keeping them at 4°C for 4 h just before challenge. The protection was more striking in mice that had not received pertussigen.

REFERENCES

1. G. S. Wilson and A. A. Miles, in *Topley and Wilson's Principles of Bacteriology, Virology and Immunity,* 6th ed., Williams and Wilkins Co., Baltimore, Md, 1975.
2. S. Malkiel and B. J. Hargis, *J. Allergy, 30,* 387 (1959).
3. M. Solotorovsky and S. Winsten, *J. Immunol., 72,* 177 (1954).
4. C. L. Fox, *J. Physiol., 192,* 241 (1958).
5. A. H. Wheeler, E. M. Brandon, and H. Patrenco, *J. Immunol., 65,* 687 (1950).
6. P. Morgan, *Proc. Soc. Exptl. Biol. Med., 102,* 161 (1959).
7. G. Eldering, *Amer. J. Hyg., 36,* 294 (1942).
8. A. G. Ospeck and M. E. Roberts, *J. Infect. Diseases, 74,* 22 (1944).
9. I. A. Parfentjev, M. A. Goodline, and M. E. Virion, *J. Bacteriol., 53,* 597 (1947).
10. S. Malkiel and B. J. Hargis, *J. Allergy, 23,* 352 (1952).
11. S. Malkiel and B. J. Hargis, *Proc. Soc. Exptl. Biol. Med., 80,* 122 (1952).
12. J. Munoz, R. F. Smith, and R. L. Cole, *Symp. Series Immunobiol. Standard.,* Vol. 13, Karger, Basel, 1970, p. 265.
13. S. Malkiel and B. J. Hargis, *J. Allergy, 35,* 306 (1964).
14. L. S. Kind, *J. Immunol., 79,* 238 (1957).
15. J. Munoz, *J. Immunol., 90,* 132 (1963).
16. F. M. Dietrich, A. Komarek, and C. Pericin, *Int. Arch. Allergy Appl. Immunol., 40,* 495 (1971).
17. R. S. Weiser, O. J. Golub, and D. M. Hamre, *J. Infect. Diseases, 68,* 97 (1941).
18. M. Solotorovsky and S. Winsten, *J. Immunol., 71,* 296 (1953).
19. J. Munoz and R. L. Anacker, *J. Immunol., 83,* 502 (1959).

20. M. Pittman and F. G. Germuth, *Proc. Soc. Exptl. Biol. Med., 87,* 425 (1954).
21. J. Munoz, L. F. Schuchardt, and W. F. Verwey, *Red. Proc., 13,* 507 (1954).
22. J. Munoz, L. F. Schuchardt, and W. F. Verwey, *J. Immunol., 80,* 77 (1958).
23. S. Tokuda, R. S. Weiser, J. Munoz, and C. Laxson, *J. Infect. Diseases, 112,* 77 (1963).
24. J. Munoz and R. L. Anacker, *J. Immunol., 83,* 640 (1959).
25. R. L. Anacker and J. Munoz, *J. Immunol., 87,* 426 (1961).
26. J. Munoz and R. K. Bergman, *Bacteriol. Rev., 32,* 103 (1968).
27. J. Munoz and L. F. Schuchardt, *J. Allergy, 25,* 125 (1954).
28. J. Munoz and M. Maung, *Proc. Soc. Exptl. Biol. Med., 106,* 70 (1961).
29. J. Munoz and R. K. Bergman, *Nature, 205,* 199 (1965).
30. R. K. Bergman and J. Munoz, *J. Immunol., 95,* 1 (1965).
31. R. K. Bergman and J. Munoz, *J. Allergy Appl. Immunol., 55,* 378 (1975).
32. M. A. Fink, *Proc. Soc. Exptl. Biol. Med., 92,* 673 (1956).
33. S. Tokuda and R. S. Weiser, *J. Immunol., 86,* 292 (1961).

Chapter 5

EFFECT OF *BORDETELLA PERTUSSIS* ON ANTIBODY PRODUCTION
AND HYPERSENSITIVITY REACTIONS

I.	GENERAL REMARKS	123
II.	ADJUVANT ACTION OF ENDOTOXIN	126
III.	ADJUVANT EFFECT OF WHOLE CELLS	127
IV.	STIMULATION OF IgE-LIKE ANTIBODY	130
V.	SUPPRESSION OF ANTIBODY FORMATION, DELAYED HYPERSENSITIVITY, AND TISSUE GRAFT REJECTION	136
VI.	MECHANISM OF ADJUVANT ACTION OF *B. PERTUSSIS*	138
	REFERENCES	139

I. GENERAL REMARKS

In 1947 Greenberg and Fleming (1,2) observed that tetanus and diphtheria toxoids given to children in combination with pertussis vaccine induced higher production of antitoxins than when the toxoids were given alone (Table 1). Their findings have since been amply confirmed by many workers (3). Pertussis vaccine acts as an adjuvant to many antigens in various animal species such as guinea pigs, mice, rats, and rabbits. Some workers have used purified protein antigens like HEA, BSA, bovine gamma globulin, horse serum albumin, chicken gamma globulin, and hemocyanin to study this adjuvant effect, while others have employed complex mixtures, such as various animal sera or erythrocytes from sheep, goats, and chickens. With all these antigens, pertussis vaccine produced a marked increase in humoral

TABLE 1

Immunization with Diphtheria Toxoid Alone and in Combination with Pertussis Vaccine[a]

Preparation	Dilution of diphtheria toxoid[b]	No. of guinea pigs	Results, % Shick negative	Efficacy compared to control
Diphtheria toxoid control	Undiluted	18	44	--
	1:2	18	33	--
	1:4	17	12	--
Diphtheria toxoid combined with tetanus toxoid	Undiluted	17	65	159% (Limits 80 to 318)
	1:2	18	41	
	1:4	18	22	
Diphtheria toxoid combined with tetanus toxoid and pertussis vaccine	1:2	18	94	915% (Limits 378 to 2,713)
	1:4	18	83	
	1:8	18	50	

[a]Data taken from Ref. 2.
[b]All the dilutions were administered s.c. in a dose of 1 ml.

antibody titers and, when studied, an increase in the number of antibody-forming cells in the spleen. The degree of stimulation was not as great as that induced by incomplete Freund's adjuvant (4)(Fig. 1) but it was as pronounced as that produced by many other adjuvants such as endotoxins and alum. In addition, pertussis vaccine has a unique and interesting adjuvant activity: it stimulates, more effectively than most other adjuvants, the production of an antibody in mice and rats which is analogous to the IgE immunoglobulin in man (5-7). Figure 2 gives results in which B. pertussis extract and incomplete Freund's adjuvant were compared for their activity to stimulate IgE-like antibody to HEA (8).

Another interesting adjuvant effect of B. pertussis, that of increasing susceptibility to autoimmune diseases, is discussed in Chapter 6.

I. GENERAL REMARKS

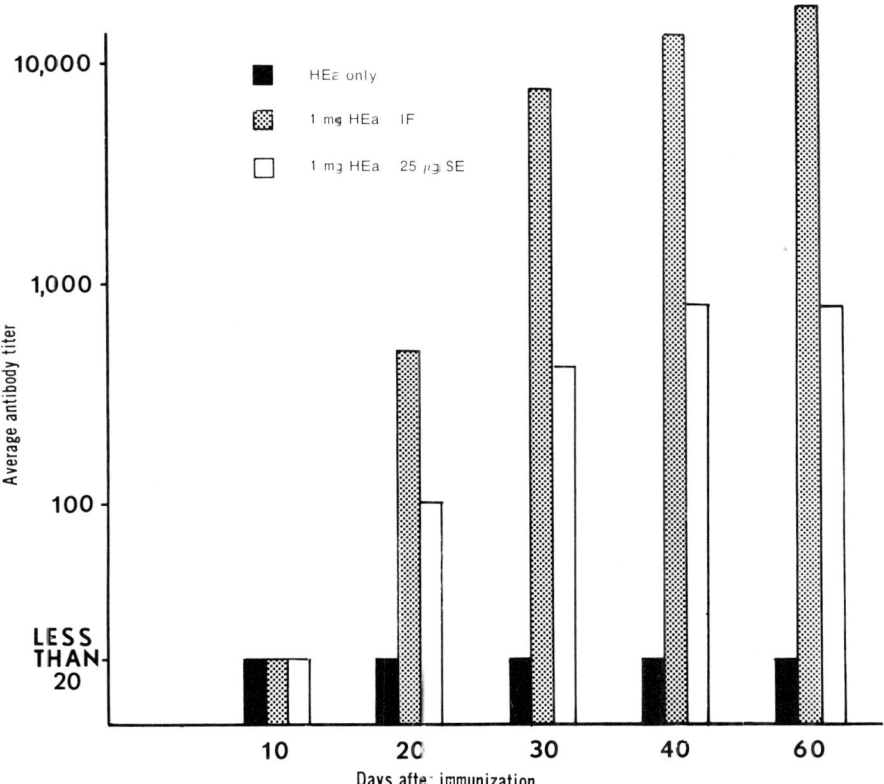

FIG. 1. Antibody response in mice inoculated with HEA. Mice were inoculated i.p. with a single 1 mg dose of HEA either alone or mixed with incomplete Freund's adjuvant or with 25 μg of saline extract from B. pertussis cells. Mice bled on day indicated. (Taken from Ref. 4.)

Whole cell vaccines are not required for the adjuvant action of B. pertussis, since soluble alkaline saline extracts from acetone-treated cells are highly effective (Fig. 2).

At least 2 distinct substances in B. pertussis cells have the ability to stimulate antibody responses. One is a heat resistant (100°C for 1 h) lipopolysaccharide (endotoxin), and the other substance is heat labile (destroyed at 80°C for ½ h). This latter adjuvant is pertussigen.

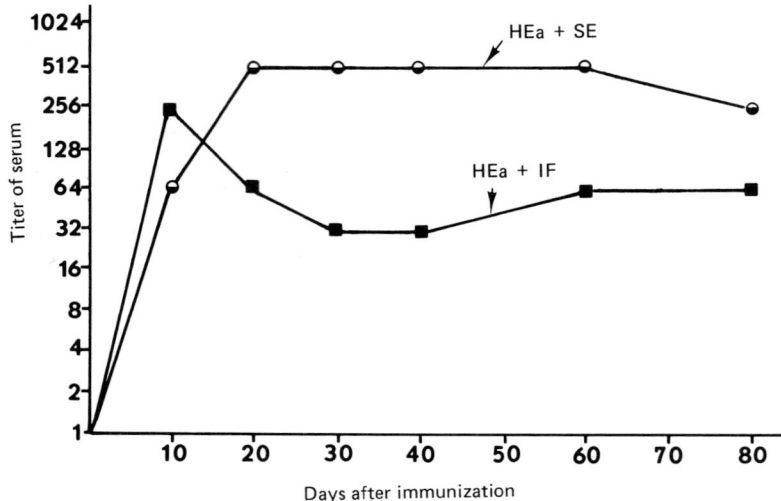

FIG. 2. 72-h PCA antibody titer (IgE) of sera from NBL/N mice following i.p. injection of 125 µg HEA and either 50 µg SE or an equal volume of Freund's incomplete adjuvant. (Taken from Ref. 8.)

The majority of the studies on the adjuvant action of *B. pertussis,* other than those performed with endotoxin, have been done with whole cells and consequently the observations have been the result of simultaneous action of at least 2 adjuvants.

II. ADJUVANT ACTION OF ENDOTOXIN

The adjuvant action of endotoxin from *B. pertussis* was first reported by Farthing and Holt (9,10). These workers found endotoxin from *B. pertussis* as prepared by the method of Westphal et al. (11), to be a good immunological adjuvant for diphtheria toxoid in the guinea pig. The effect was demonstrated by earlier production and a higher antitoxin serum titer for at least 4 weeks after immunization. The adjuvant effect occurred only with the primary stimulus. Farthing (9) also prepared lipid A from *B. pertussis* endotoxin by the method of Westphal and Lüderitz (12) and found it to be active as an adjuvant in doses greater than those for endotoxin; however, even at the high doses, it did not equal the response obtained with intact endotoxin (Fig. 3).

III. ADJUVANT EFFECT OF WHOLE CELLS

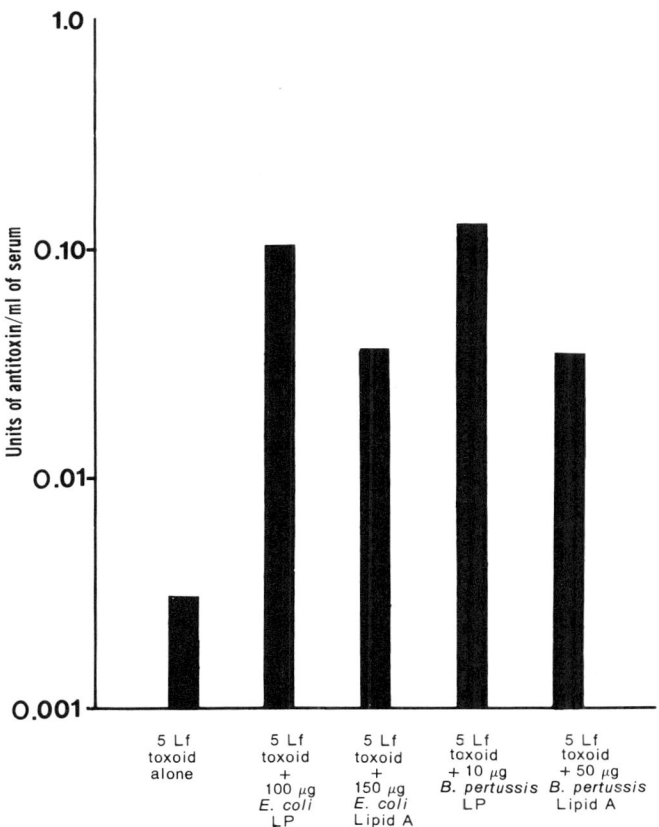

FIG. 3. Adjuvant action of *E. coli* and *B. pertussis* endotoxins and their lipid A components. (Taken from Ref. 9.)

III. ADJUVANT EFFECT OF WHOLE CELLS

The adjuvant effect of whole-cell pertussis vaccine on antibody-forming cells has been extensively studied by Finger and co-workers in Germany (13). They have found a marked stimulation in the primary response to 4×10^8 sheep erythrocytes given i.p. when about 4×10^9 killed *B. pertussis* cells were given simultaneously. The humoral antibody titers were elevated and the number of plaque-forming cells (PFC) in the spleen was increased. In addition, an accelerated and prolonged multiplication of both 19 S and 7 S hemolysin-producing spleen cells was induced by the vaccine (14). It was most effective when given mixed with the antigen, but it also was somewhat effective when given up to 48 h before antigen. As time between injection of pertussis vaccine and antigen was increased, the adjuvant effect decreased. Dresser et al. (15) found a slight reduction of PFC when the vaccine was given 3 or 1 day before sheep erythrocytes. The adjuvant effect was most pronounced in the IgG_{2b} antibody. When *B. pertussis* was given together with the second antigenic stimulus 41 days after the primary immunization with 4×10^8 erythrocytes the peak values of indirect spleen PFC did not differ significantly during the first 7 days from those of corresponding controls receiving the booster of erythrocytes only, but at 10 to 21 days after the booster there was a detectable enhancement effect of *B. pertussis* on the PFC (14). The adjuvant effect on secondary response was detected when *B. pertussis* was given simultaneously with the antigen or 12, 24, or 48 h before the sheep erythrocytes. On secondary stimulation, pertussis vaccine increases to a greater extent the IgG PFC (14), but an effect is also observed on the IgM-forming cells (14,16). The increase in PFC was detected as early as 30 h after immunization and could still be detected 35 days after a single immunization (16). Although endotoxin was undoubtedly present, Finger et al. (16) felt that it was not entirely responsible for this effect. Endotoxin causes an accelerated and prolonged development of direct (IgM) and indirect (IgG) PFC in the spleen and also increases the priming effect for a secondary response (17). These effects are similar to those found for *B. pertussis* cells and most of the effects observed could have been due to the endotoxin content of the cells. It has been found, however, that the adjuvant effect of *B. pertussis* can still be shown in endotoxin tolerant mice (18) and we (unpublished observations) have shown that the activity of *B. pertussis* extracts at the low doses employed is mostly destroyed by heating at 80°C for ½ h (Fig. 4).

III. ADJUVANT EFFECT OF WHOLE CELLS

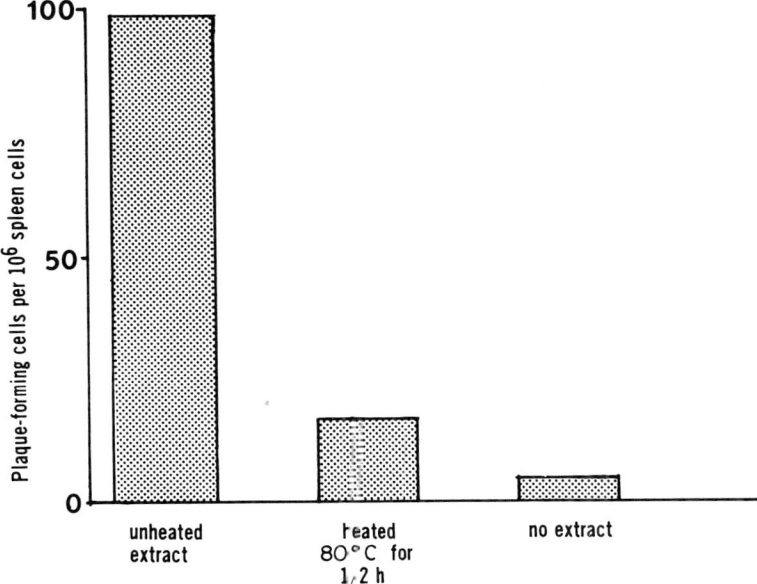

FIG. 4. Effect of heat on ability of *B. pertussis* extract to stimulate IgG antibody-forming cells in spleens of mice. Mice received i.p. 0.8 ml of a mixture containing 50 μg of chicken gamma globulin and 0.2 M fraction (40 SD_{50}'s of HSF). The 1st bar gives the results with unheated 0.2 M fraction, the 2nd with the same material heated at 80°C for ½ h, and the 3rd the number of plaques in control immunized mice. (Data from R. K. Bergman and J. J. Munoz, unpublished observations.)

It is interesting that pertussis vaccine alone increases the IgM PFC to sheep erythrocytes (15), thus complicating the studies employing this antigen. In nude, thymusless mice, *B. pertussis* stimulates only the direct PFC (IgM producers). In these animals the vaccine produces splenomegaly and leukocytosis entirely due to granulocytes, while in normal mice lymphocytosis is also produced (19).

Unanue et al. (20) found that macrophages with engulfed antigen (hemocyanin), when treated in vitro with pertussis vaccine, were more effective in stimulating antibody production when given to a mouse than were an equal number of untreated macrophages. The adjuvant effect of *B. pertussis* was exerted only after these cells were taken up by the macrophages. Furthermore, the adjuvant effect was also observed when adjuvant-containing macrophages (without hemocyanin) were added to macrophages containing hemocyanin. They concluded that adjuvant action in this system did not require the presence of antigen and adjuvant in the same cell.

IV. STIMULATION OF IgE-LIKE ANTIBODY

One of the most interesting adjuvant actions of *B. pertussis* is its ability to stimulate IgE in mice and rats. This effect is mainly due to the heat labile adjuvant. This action was first described by Mota (5-7) when he showed that pertussis vaccine stimulated the production of so-called "mast cell sensitizing antibody" to antigens given with it. This antibody was characterized by (a) its ability to fix to mast cells and degranulate them in the presence of the specific antigen; (b) losing its sensitizing ability when heated at 56°C for 1 to 3 h or when treated with 2-mercaptoethanol; (c) fixing to skin site for a long period of time (over 72 h); and (d) migrating in agar gel electrophoresis toward the cathode. This immunoglobulin was eventually shown to be a unique protein distinct from other mouse immunoglobulins (21). Our experience (4,8) with the induction of this immunoglobulin by *B. pertussis* has confirmed that the 72-h PCA antibody has properties similar to those of IgE of man. Since the discovery of this activity of *B. pertussis,* numerous investigators have used pertussis vaccine to stimulate the production of IgE-like antibody in animals.

The active principle is pertussigen, since highly purified fractions induce the production of IgE in mice (see Chapter 2 for purification details). With some of these preparations, as little as 0.075 µg (expressed as protein) mixed with 125 µg HEA has stimulated in CFW mice IgE and IgG_1 production with specificity to HEA (Table 2). The actual amount of each antibody is not known, but since IgE

IV. STIMULATION OF IgE-LIKE ANTIBODY

TABLE 2

Stimulation of 2-h (IgG_1) and 72-h (IgE) Antibody by a Highly Purified Preparation of Pertussigen[a,b]

Amount	Titer			
	2-h[d]		72-h	
μg of protein[c]	Pre[e]	Post[f]	Pre	Post
0.75	100	>1,000	100	1,000
0.075	10	1,000	10	1,000
0.0075	<10	100	<10	100
0	<10	100	<10	100

[a]Data from R. K. Bergman and J. J. Munoz, unpublished observations.
[b]Fraction B obtained by sucrose density gradient electrophoresis as described in Chapter 2 was used. Mice received i.p. 125 μg of HEA mixed with appropriate amount of fraction in 0.2 ml final volume.
[c]Determined by Folin-Ciocalteau method.
[d]The sera were heated at 56°C for 3 h before testing for 2-h PCA.
[e]Pre = bleeding on day 28 before booster.
[f]Post = bleeding on day 36, 8 days after booster dose of 5 μg HEA.

is active in very small concentrations [about 10^{-6} μg N (22)], the concentration in the blood of immunized animals must be very low, judging from the PCA antibody titers obtained. Attempts to purify and characterize IgE in mouse sera have been difficult but Prouvost-Danon et al. (21) succeeded in establishing its immunological uniqueness.

Endotoxin from B. pertussis is not very active in stimulating IgE when compared to pertussigen. Salmonella minnesota endotoxin was somewhat more effective, but still inferior to B. pertussis extracts (Fig. 5).

Stimulation of IgE by B. pertussis after 1 immunizing injection follows a curve shown in Fig. 2. This response is different from that obtained with Freund's incomplete adjuvant, where a more rapid decrease in titer was observed. It has been stated that booster

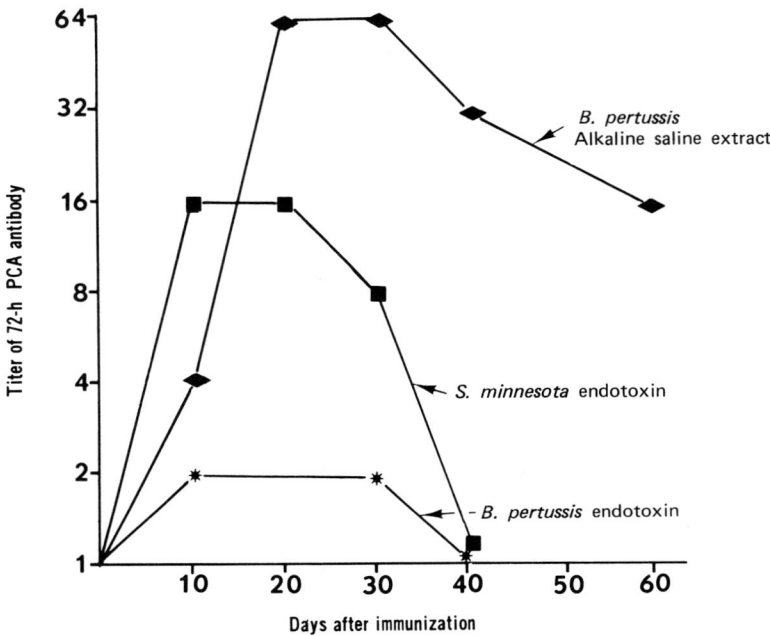

FIG. 5. Effect of *B. pertussis* endotoxin and *S. minnesota* endotoxin on IgE stimulation to HEA. Mice received 125 μg HEA mixed with 25 μg of endotoxin or 50 μg of alkaline saline extract from *B. pertussis*. (Reprinted from Ref. 4, p. 774, by courtesy of Williams and Wilkins Co., copyright 1969.)

doses of HEA, following a primary immunization with *B. pertussis* vaccine and HEA, decrease IgE production (23) but we, employing *B. pertussis* extracts as adjuvant, found that a small dose of HEA given 20 to 30 days after a primary dose induced a booster effect on IgE titers (Fig. 6), and further that 5 μg booster doses maintain the titer at a constant level.

One method employed to produce large amounts of peritoneal fluid containing IgE is to repeatedly inoculate mice i.p. with Freund's complete adjuvant and antigen after the original inoculation with *B. pertussis* extracts and HEA. With this method, large volumes of ascites (Fig. 7), containing 72-h PCA antibody to titers of as much as 1,000, are produced.

IV. STIMULATION OF IgE-LIKE ANTIBODY

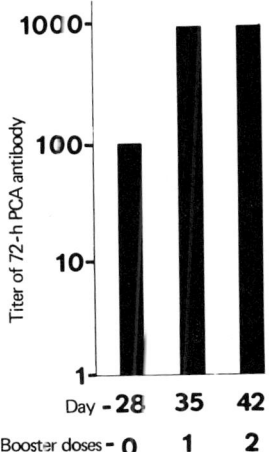

FIG. 6. Effect of small booster doses of antigen on 72-h PCA antibody in mice. C57BL/6J mice received a primary i.p. dose of 125 µg of HEA plus 50 µg of alkaline saline extract of B. pertussis on day 0. On day 28 they were bled and given an i.p. booster dose of 5 µg HEA. On day 35 they were bled again and a 2nd i.p. booster dose of 5 µg HEA given. The mice were bled again on day 42. (Data from R. K. Bergman and J. J. Munoz, unpublished observations.)

The most effective routes for stimulating increased serum levels of IgE to HEA were the i.v. and i.p. routes; the s.c. route was inferior (Fig. 8). The method of choice is to give the first dose i.p. (125 µg HEA mixed with pertussigen) and 28 days later a s.c. booster of 5 µg HEA in saline, and the mice bled 7 days later. A highly efficient booster effect can also be obtained by giving i.p. 5 µg of HEA mixed in complete Freund's adjuvant; titers of 72-h PCA antibody up to 2,500 are obtained in CFW mice. A 2nd booster given 7 days after the 1st booster did not increase the titers.

IgE with specificity to antigens found in B. pertussis extracts are also detected in mice receiving impure preparations of pertussigen.

Stimulation of IgE at the antibody-forming cell level has not been studied because no indicator to differentiate IgG from IgE-producing cells has been developed.

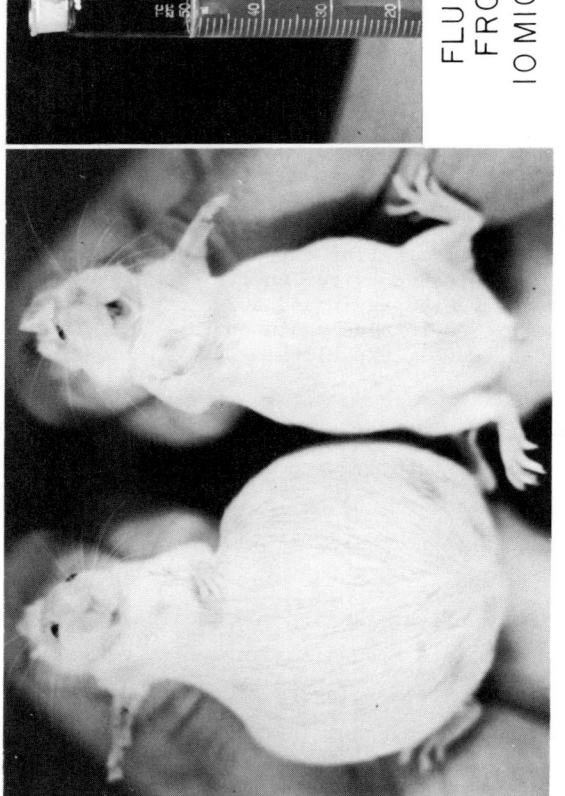

FIG. 7. Stimulation of ascitic fluid by repeated injections of Freund's complete adjuvant in mice that had received an original injection of HEA plus *B. pertussis* extract. Mouse on the left responded with large accumulation of fluid; mouse identically treated on the right did not. (Data from R. K. Bergman and J. J. Munoz, unpublished observations.)

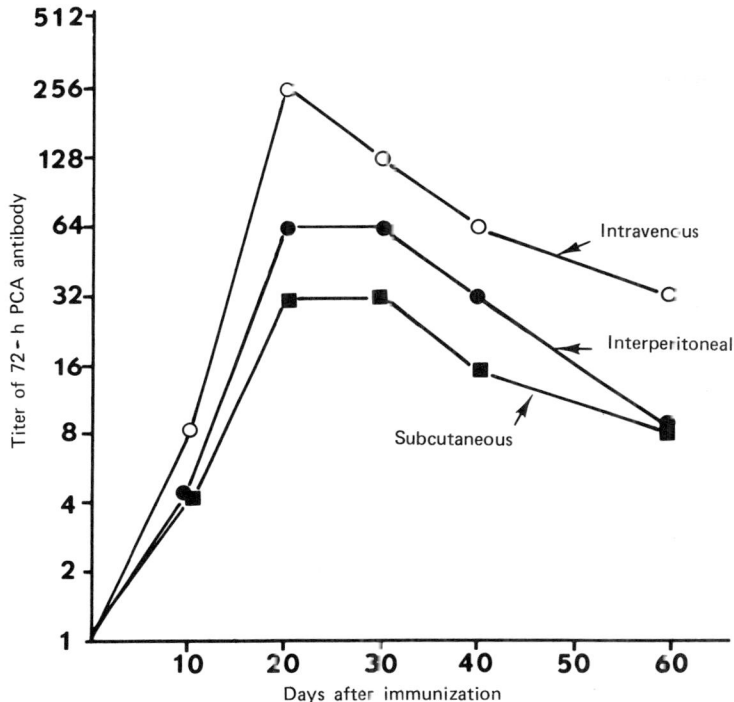

FIG. 8. Effect of route of immunization on stimulation of IgE in RML mice. Mice received i.p. 125 µg HEA and another injection of 50 µg SE given either s.c., i.p., or i.v. (Reprinted from Ref. 4, p. 773, by courtesy of Williams and Wilkins Co., copyright 1969.)

V. SUPPRESSION OF ANTIBODY FORMATION, DELAYED HYPERSENSITIVITY, AND TISSUE GRAFT REJECTION

Investigators at the National Institute of Health in Tokyo have purified LPF (pertussigen) (24) and have made some interesting observations regarding its effects on antibody production and induction of delayed hypersensitivity. Pertussigen, as purified by them, when mixed with dinitrophenylated Ascaris suum extract (25) was capable of stimulating IgE antibody in rats, as was indicated in Section IV. IgE was still increased even when pertussigen was given i.v. 15 days before antigen (25). However, the IgG and IgM response to sheep erythrocytes and tetanus toxoid was inhibited when the preparation was given 1, 2, or 3 days before antigen (26) (Fig. 9). This effect

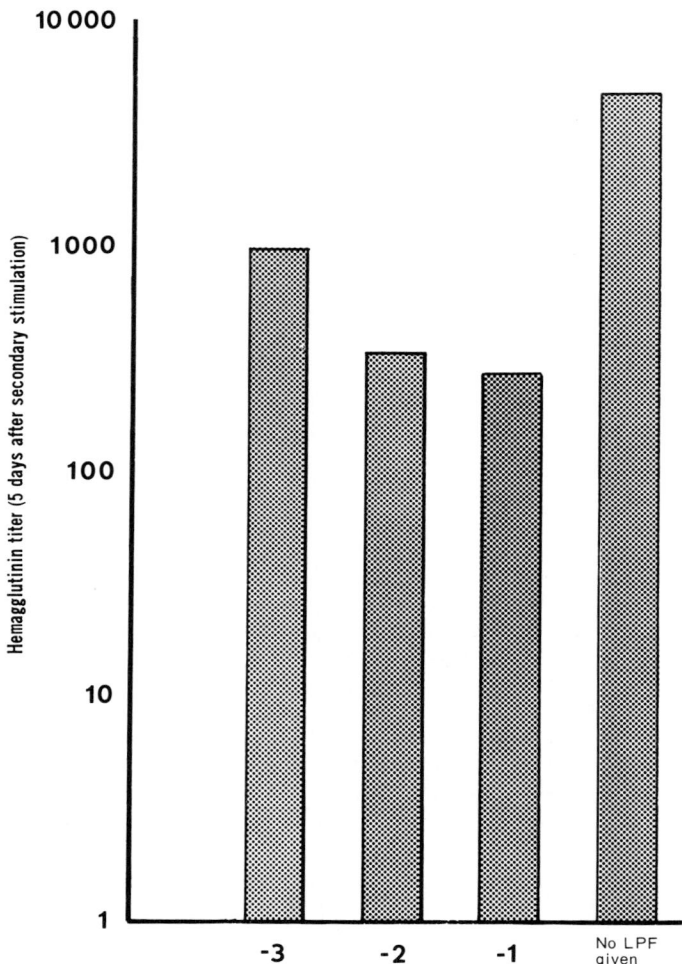

FIG. 9. Pertussigen-induced inhibition of secondary response in mice to sheep erythrocytes. Mice received pertussigen at various time intervals before primary antigenic stimulation with 1×10^7 sheep erythrocytes. Secondary antigenic stimulation (1×10^8 cells) was given 34 days after the primary inoculation. Hemagglutinin titers were determined 5 days later. (Taken from Ref. 26.)

V. SUPPRESSION OF ANTIBODY FORMATION

was more pronounced with smaller doses of antigen than with larger doses, and both primary and secondary responses were suppressed. Since many workers have found an increased antibody response when pertussis vaccine or crude extracts are given mixed with or before antigen, these observations should be thoroughly reinvestigated. Presumably these preparations were relatively free of endotoxin and it is possible that when endotoxin is removed from a B. pertussis extract its action may differ. Another very interesting observation made by Ochiai et al. (27) is that pertussigen markedly suppressed various expressions of delayed hypersensitivity. Thus, if 0.5 µg of pertussigen were given into the 4 footpads of previously sensitized rats 3 days before skin testing with 10 µg of PPD, a complete inhibition of the delayed response was observed (Table 3). Suppression was also observed if pertussigen was given 3 days before immunization

TABLE 3

Effect of Pertussigen on the Development of Delayed-Type Skin Reaction with Tuberculin in the Rat[a,b]

Treatment	Number of animals	Diameter of skin reaction (mm)	WBC count at time of skin test (cells/mm^3 blood)
None	7	10.0, 9.0, 9.0, 8.5 7.5, 7.0, 6.0	6,700
LPF 3 days before immunization	7	6.5, 5.0, 4.0, 3.0 0, 0, 0	13,750
LPF 3 days before challenge	5	0, 0, 0, 0, 0	30,570

[a]Reprinted from Ref. 27, p. 201, by courtesy of S. Karger AG, Basel.
[b]Rats were sensitized by footpad injection with 0.2 ml of Freund's complete adjuvant emulsified with an equal volume of saline. The rats were skin-tested 14 days later with 10 µg PPD. One group of rats received no further treatment, another received 0.5 µg LPF distributed in all 4 footpads 3 days before sensitization, and the last group received the same amount of LPF 3 days before skin testing with PPD.

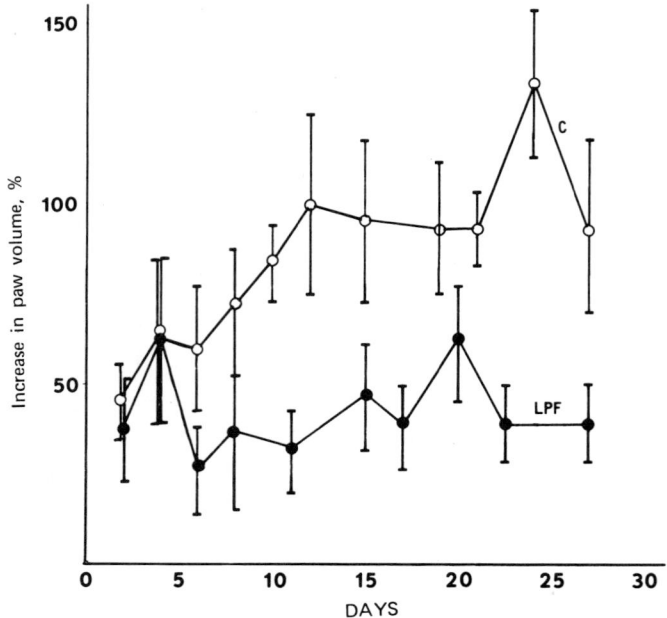

FIG. 10. Effect of pertussigen on the development of adjuvant disease in the rat. Mean swelling of the paw volume of groups of 7 pertussigen-treated (black circles) and control (open circles) rats is plotted. Vertical bars represent standard deviations. (Reprinted from Ref. 27, p. 202, by courtesy of S. Karger, Basel.)

with complete Freund's adjuvant. Adjuvant arthritis in rats was also suppressed when pertussigen was given in the footpads 4 days before Freund's adjuvant (Fig. 10); furthermore, rejection of skin grafts were delayed by a single i.v. injection of 0.1 µg pertussigen given to mice 4 days before or simultaneously with the graft. Similar suppression of skin graft rejection was obtained by Ptak et al. employing pertussis vaccine (28).

VI. MECHANISM OF ADJUVANT ACTION OF *B. PERTUSSIS*

Available information does not clearly explain the action of *B. pertussis* cells in stimulating the antibody response. Whole cell vaccines contain endotoxin as well as pertussigen, both of which act as adjuvants but not necessarily by similar mechanisms. The mode of action of endotoxins from other Gram-negative bacteria has been investigated by many workers and those interested can find many papers and reviews on the topic (29). Endotoxin from B. pertussis must act similarly and the studies which are discussed below undoubtedly were concerned, at least partly, with the action of endotoxin. The mode of action of pertussigen has not yet been carefully studied.

According to Finger et al. (13) B. pertussis cells induce increased recruitment of antigen-responding cells and proliferation of memory cells. Reed et al. (30) also concluded that B. pertussis exerted its effect directly on the precursor of antibody-forming cells by increasing their rate of cell division. The adjuvant effect is greater on the IgG class of immunoglobulins than on the other Ig classes (15,31,32), and within the IgG classes the order of increase is $IgG_1 > IgG_{2b} > IgG_{2a}$. IgA is increased to a greater extent than IgM (31). At low concentrations of antigen, the adjuvant effect is dependent on T cells and the IgG_1 is more dependent on these cells than is the IgM response (33-35). The adjuvant effect on the IgE response seems to be totally dependent on T cells, because thymusless mice do not respond with the production of IgE (J. J. Munoz and N. W. Reed, unpublished observations).

Various workers (14,36) have demonstrated an adjuvant action of B. pertussis cells when given as early as 4 days before antigen. With purified pertussigen some (26,27) have shown that when it is given 1, 2, or 3 days before antigen, an inhibition of antibody production and of delayed hypersensitivity response is observed. To elucidate the mechanism of adjuvant action of pertussigen, investigations with pure preparations are needed.

REFERENCES

1. L. Greenberg and D. S. Fleming, Can. J. Public Health, 38, 279 (1947).
2. L. Greenberg and D. S. Fleming, Can. J. Public Health, 39, 131 (1948).

3. J. Munoz, *Adv. Immunol., 4,* 397 (1964).
4. C. R. Clausen, J. Munoz, and R. K. Bergman, *J. Immunol., 103,* 768 (1969).
5. I. Mota, *Nature, 182,* 1021 (1958).
6. I. Mota, *Life Sci., 7,* 465 (1963).
7. I. Mota and J. M. Peixoto, *Life Sci., 5,* 1723 (1966).
8. C. R. Clausen, J. Munoz, and R. K. Bergman, *J. Immunol., 104,* 312 (1970).
9. J. R. Farthing, *Brit. J. Exptl. Pathol., 42,* 614 (1961).
10. J. R. Farthing and L. B. Holt, *J. Hyg., 60,* 411 (1962).
11. O. Westphal, O. Lüderitz, and F. Bister, *Z. Naturforsch., 7b,* 148 (1952).
12. O. Westphal and O. Lüderitz, *Angew. Chem., 66,* 407 (1954).
13. H. Finger, M. Bartoschek, and P. Emmerling, *Infect. Immun., 2,* 590 (1970).
14. H. Finger, P. Emmerling, and E. Brüss, *Infect. Immun., 1,* 251 (1970).
15. D. W. Dresser, H. H. Wortis, and H. R. Anderson, *Clin. Exptl. Immunol., 7,* 817 (1970).
16. H. Finger, P. Emmerling, and H. Schmidt, *Experientia, 23,* 591 (1967).
17. H. Finger, G. Beneke, and H. Fresenius, *Int. Arch. Allergy Appl. Immunol., 38,* 598 (1970).
18. H. Finger, I. Fölmer, L. Plager, and M. Henseling, *Experientia, 28,* 562 (1972).
19. H. Finger, W. Mohr, H. Hof, E. Elekes, and L. Plager, *Med. Microbiol. Immunol., 159,* 13 (1973).
20. E. R. Unanue, B. A. Askonas, and A. C. Allison, *J. Immunol., 103,* 71 (1969).
21. A. Prouvost-Danon, R. Binaghi, S. Rochas, and Y. Boussac-Aron, *Immunology, 23,* 481 (1972).
22. I. Ishizaka and T. Ishizaka, *J. Allergy, 42,* 330 (1968).
23. I. Mota, *Immunology, 7,* 681 (1964).
24. S. Iwasa, S. Ishida, S. Asakawa, and M. Kurokawa, *Jap. J. Med. Sci. Biol., 21,* 363 (1968).
25. T. Tada, K. Okumura, T. Ochiai, and S. Iwasa, *Int. Arch. Allergy, 43,* 207 (1972).
26. S. Asakawa, *Jap. J. Med. Sci. Biol., 22,* 23 (1969).

REFERENCES

27. T. Ochiai, K. Okumura, T. Tada, and S. Iwasa, *Int. Arch. Allergy, 43,* 196 (1972).
28. W. Ptak, H. Festenstein, G. L. Asherson, and P. M. Denman, *Nature, 222,* 1083 (1969).
29. J. Munoz, *Adv. Immunol., 4,* 397 (1964).
30. C. E. Reed, M. Benner, S. D. Lockey, T. Enta, S. Makino, and R. H. Carr, *J. Allergy Clin. Immunol., 49,* 174 (1972).
31. D. W. Dresser and J. M. Phillips, in *Immunopotentiation,* Ciba Found. Symp. 18 (New Series, Elsevier, 1973), pp. 3-28.
32. G. Torrigiani, *Clin. Exp. Immunol., 11,* 125 (1972).
33. R. B. Taylor and H. H. Wortis, *Nature, 220,* 927 (1968).
34. D. W. Dresser, *Eur. J. Immunol., 2,* 50 (1972).
35. R. N. Taub and R. K. Gershon, *J. Immunol., 108,* 377 (1972).
36. D. W. Dresser, *Clin. Exp. Immunol., 3,* 877 (1968).

Chapter 6

EFFECT ON AUTOIMMUNE DISEASES

I. GENERAL REMARKS 143
II. HYPERACUTE EXPERIMENTAL ALLERGIC ENCEPHALOMYELITIS. . . . 145
III. OTHER AUTOIMMUNE DISEASES 148
IV. MECHANISM OF ACTION 148
    REFERENCES. 150

I. GENERAL REMARKS

Another expression of the immunological adjuvant action of B. pertussis is the ability to accelerate the autoimmune disease process and to render various animal species more susceptible to this type of diseases. In most of these diseases it is not known whether circulating antibodies are involved or if, as it is generally thought, the disease is mediated entirely by cellular hypersensitivity in which circulating antibodies play little or no part. In some autoimmune diseases in which B. pertussis has a pronounced effect, its action seems to involve other changes, in addition to increased antibody response or cellular hypersensitivity. Physiological changes such as increased permeability of the capillaries may be important.

The first to show that pertussis vaccine enhanced development of an autoimmune disease in animals were Lee and Olitsky (1) who showed that pertussis vaccine given concomitantly with spinal cord

TABLE 1

Clinical Signs of EAE in Inbred and Outbred Mice[a,b]

Strain	Inoculated in 4 feet		Inoculated in cervical nodes	
	i.v. pertussis	No pertussis	i.v. pertussis	No pertussis
SJL/J	9/10[c]	0/6	4/7	0/5
SWR/J	13/17	1/13		0/4
C57L/J	11/16	0/5	12/13	6/10
CBA/J	1/21	0/5	12/16	1/5
C57BL/6J	2/6			
A/J	1/6			
129/J	1/6			
AKR/J	1/13		0/7	0/5
RF/J	0/6			
Balb/cJ	0/16		0/7	0/5
DBA/2J	0/4			
ST/bJ	0/3			
BRVR	0/12			
BSVS	0/10			
Camm SW[d]	19/35	0/5	3/5	1/5
CF#1[d]	4/6	0/6		
CFW[d]	3/7	0/6		
TAC/SW[d]	0/6			

[a] Reprinted from Ref. 2, p. 141, by courtesy of Williams and Wilkins Co., copyright 1973.

[b] Mouse cord-Freund's adjuvant emulsion used throughout.

[c] Numerator = number of mice with clinical signs, denominator = total number of mice. Denominators of 9 or more represent combined results of 2 or 3 trials. All clinical signs were confirmed by abundant histologic lesions.

[d] Outbred strains. All other strains are inbred.

emulsions increased the incidence of experimental allergic encephalomyelitis (EAE) in mice. Many workers have found it difficult to induce EAE in mice, but recently Levine and Sowinski (2) and Bernard and Carnegie (3) have done so with certain mouse strains. EAE was induced in 3 of 14 inbred strains which received by footpad injection mouse spinal cord emulsified in complete Freund's adjuvant and a concurrent i.v. injection of pertussis vaccine. Within 15 to 20 days, mice of susceptible strains developed symptoms and histological lesions of EAE. Without pertussis vaccine, symptoms of EAE were rarely noted (Table 1).

II. HYPERACUTE EXPERIMENTAL ALLERGIC ENCEPHALOMYELITIS

Most research on the effect of whole cell pertussis vaccines on induction of EAE has been done in rats (4-6). Pretreatment with pertussis vaccine 3 to 4 days before injection of spinal cord emulsion was shown to make some strains of rats more susceptible to EAE (7). Pertussis vaccine also can substitute for mycobacteria in Freund's adjuvant for inducing EAE in the guinea pig (8). The most interesting observation regarding the effect of pertussis vaccine, however, is its effect on the type of EAE produced by spinal cord emulsions in Lewis Rats (9). This strain of rats is mildly susceptible to EAE, even when spinal cord emulsions are given without an adjuvant. The form of EAE developed under these conditions or when Freund's adjuvant is used is of the ordinary type with an onset at 10 to 14 days and a typical mononuclear perivascular infiltration in the spinal cord. When pertussis vaccine is given i.p. in a small dose with the antigen (guinea pig spinal cord), Lewis rats develop symptoms of EAE as early as the 6th day (Table 2). Furthermore, the histological changes in the cord are different, because the cellular infiltrate contains, in addition to mononuclear cells, many neutrophils and fibrin. This histological change is seldom seen in the

TABLE 2

Titration of Pertussis Vaccine for Production of Hyperacute Form of EAE[a]

Dose[b] (ml)	EAE onset[c]	Lesions with fibrin[d]
0.12	8, 8, 8, 8, 8	4, 4, 4, 3, 4
0.06	6, 6, 7, 7	4, 2, 4, 3
0.03	7, 7, 7, 7, 7	4, 4, 3, 3, 0
0.015	6, 7, 7, 8, 12	1, 1, 1, 3, 0
0.0075	7, 7, 7, 7, 8	1, 0, 0, 0, 1
None	9, -, -, -, -	0, 0, 0, 0, 0

[a]Reprinted from Ref. 5, p. 365, by courtesy of Williams and Wilkins Co., copyright 1966.

[b]Stated dose of pertussis vaccine concentrate (200 billion organisms/ml; 4 mg dry weight/ml) combined with 200 mg (wet weight) guinea pig cord homogenate, diluted to 3 ml with saline, injected i.p.

[c]Each number represents days of clinical onset of EAE in an individual rat; each hyphen indicates a rat with no clinical signs (but all rats had histologic evidence of EAE).

[d]Graded from 0 to 4 according to number of spinal cord vessels surrounded by fibrin.

ordinary form of EAE (9). Levine called this accelerated form of the disease hyperacute EAE. This form of EAE is produced only when the right dose and type of antigen are used. Furthermore, the strain of rat and route of inoculation are critical. When any of these conditions are not met the disease produced, if any, is ordinary EAE. It is interesting that passive transfer of lymph node cells from donor rats with hyperacute EAE produces only ordinary EAE in normal rats. This observations shows 2 important facts: (a) hyperacute EAE is only a form of ordinary EAE and (b) pertussis vaccine does not act by changing the lymph node cells. This vaccine may act, as is discussed later, by increasing vascular permeability of the blood-brain barrier. Pertussis vaccine is effective when mixed with antigen and given in the footpad or when given at a separate site but in

II. HYPERACUTE EAE

close proximity to the antigen. It is effective when given i.v. but not as effective if given in the opposite foot, in the neck region, or in the peritoneal cavity. The vaccine is also effective when given after the antigen or sometime before (10). In passive transfer experiments, lymphoid cells from donors with ordinary EAE produce in rats pretreated with pertussis vaccine hyperacute EAE (10). This again suggests that a fundamental change has occurred in the rat treated with pertussis vaccine and not in the lymphoid cells.

The active factor in pertussis vaccine is, according to Levine et al. (5), HSF because (a) it is heat labile (80°C for ½h), (b) semi-purified preparations of HSF were active, (c) preparations treated with 1% formaldehyde were inactive, and (d) during fractionation the

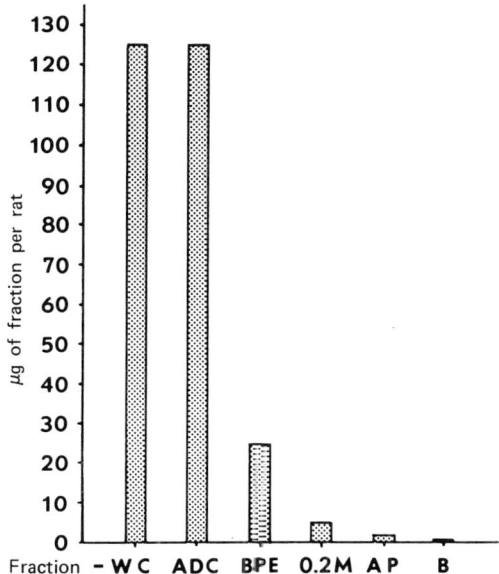

FIG. 1. Amount of various active fractions from B. pertussis required to induce hyperacute EAE. WC = washed and lyophilized whole cells, ADC = acetone-extracted cells, BPE = alkaline saline extract dialyzed and lyophilized, 0.2 M = 0.2 M buffer fraction from hydroxyl-apatite column, AP = pH 4.5 precipitate obtained from 0.2 M fraction, B = active fraction from sucrose density gradient electrophoresis of AP.

potency closely follows HSF activity. We have tested crude and purified pertussigen preparations for their ability to induce hyperacute AEA, and found that the purest preparations were highly active (11). Less than 5 µg of purified preparations of pertussigen mixed with 200 mg (wet weight) of guinea pig spinal cord emulsion induced EAE (Fig. 1).

III. OTHER AUTOIMMUNE DISEASES

Pertussis vaccine also enhances the production of aspermatogenesis in some strains of mice. In the CD-1 mouse, Malkiel et al. (12,13) produced this disease by giving 4 mg (wet weight) of homologous testicular tissue mixed with 6×10^9 B. pertussis cells into each hind footpad, followed by a similar dose given i.p. 17 days later. Twenty days after the 2nd injection, marked disorganization and disintegration of the tubular elements occurred. Characteristic multinucleated cells also appeared, and antibodies and delayed skin type reaction to testes extracts were demonstrated.

Thyroiditis has also resulted after rate received B. pertussis vaccine plus thyroid tissue emulsions (14). Soluble fractions have also been used with limited success (N. Rose, personal communication). To a lesser degree pertussis vaccine has also helped in the induction of adrenalitis in mice (15).

IV. MECHANISM OF ACTION

The mechanism by which B. pertussis increases the susceptibility of animals to autoimmune disease is not entirely clear. The increased production of antibodies to encephalitogenic antigen by pertussigen does not seem to explain satisfactorily the acceleration of EAE in Lewis rats. If this were the case, one would expect that a more powerful adjuvant, such as complete Freund's adjuvant (16,17) given with the spinal cord suspension, would also induce an accelerated EAE. Pertussigen, we believe, has a dual action: One is the stimulation of antibodies and/or cellular hypersensitivity to the enceph-

IV. MECHANISM OF ACTION

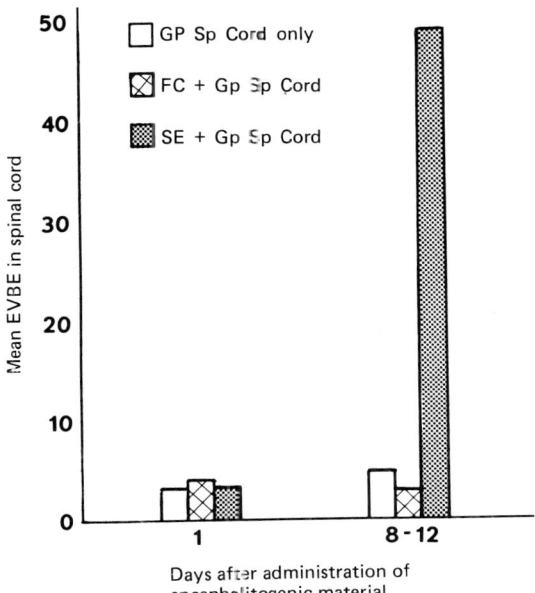

FIG. 2. Permeability in Lewis rat spinal cords following administration of encephalitogenic material. One group of rats received 200 mg (wet weight) guinea pig spinal cord (Gp Sp Cord) emulsion i.p. in 4 ml saline. A 2nd group received 200 mg of spinal cord in an emulsion of 2 ml saline + 2 ml Freund's complete (FC) adjuvant given i.p. The 3rd group received 200 mg spinal cord and 200 µg of saline extract (SE) from B. pertussis in 4 ml saline given i.p. Permeability of the spinal cord is expressed as extravascular blood equivalent (EVBE) after the method of Leibowitz and Kennedy (18) (see Appendix). Each rat received 5 µCi [^{131}I]HSA and 5 µCi [^{125}I]HSA in the permeability test.

alitogenic antigen, and the other is an increase in the vascular permeability in the spinal cord. These effects may be further reinforced by the induction of an increased sensitivity to vasoactive amines (histamine, serotonin, etc.) which are released when an antigen-antibody reaction takes place in vivo. In the case of EAE, this release would occur in the central nervous system where the encephalitogenic antigen is found. Release of vasoactive amine in the proximity of spinal cord capillaries would increase the permeability of these vessels disproportionately in the pertussigen-treated animal (Fig. 2). Because of the apparent adrenergic blocking effect of pertussigen (19), epinephrine cannot counteract the effects of vasoactive substances.

If lymphoid cells are responsible for EAE, rather than humoral antibodies to the encephalitogenic antigen, the increased permeability of capillaries at the blood-brain barrier could also facilitate the migration of these cells into the central nervous tissue and thus produce accelerated induction of EAE.

REFERENCES

1. J. M. Lee and P. K. Olitsky, *Proc. Soc. Exptl. Biol. Med.*, *89*, 263 (1955).
2. S. Levine and R. Sowinski, *J. Immunol.*, *110*, 139 (1973).
3. C. A. Bernard and P. R. Carnegie, *J. Immunol.*, *114*, 1537 (1975).
4. S. Levine and E. J. Wenk, *Ann. N. Y. Acad. Sci.*, *122*, 209 (1965).
5. S. Levine, E.J. Wenk, H. B. Devlin, R. E. Pieroni, and L. Levine, *J. Immunol.*, *97*, 363 (1966).
6. S. Levine and R. Sowinski, *Amer. J. Pathol.*, *73*, 247 (1973).
7. S. Levine and E. J. Wenk, *Ann. N. Y. Acad. Sci.*, *122*, 209 (1965).
8. S. L. Weiner, M. Tinker, and W. L. Bradford, *AMA Arch. Pathol.*, *67*, 694 (1959).
9. S. Levine and E. J. Wenk, *Amer. J. Pathol.*, *47*, 61 (1965).
10. S. Levine and E. J. Wenk, *Amer. J. Pathol.*, *50*, 465 (1967).
11. R. K. Bergman and J. J. Munoz (in press).
12. B. J. Hargis, S. Malkiel, and J. Berkelhammer, *J. Immunol.*, *101*, 374 (1968).
13. S. Malkiel, B. J. Hargis, and J. Berkelhammer, *J. Allergy*, *46*, 321 (1970).
14. F. J. Twarog and N. R. Rose, *Proc. Soc. Exptl. Biol. Med.*, *130*, 434 (1969).
15. S. Levine, *Science*, *158*, 1190 (1967).
16. J. Freund, *Ann. Rev. Microbiol.*, *1*, 291 (1947).
17. J. Munoz, *J. Immunol.*, *90*, 132 (1963).
18. S. Leibowitz and L. Kennedy, *Immunology*, *22*, 859 (1972).
19. C. W. Fishel and A. Szentivanyi, *J. Allergy*, *34*, 439 (1963).

Chapter 7

LYMPHOCYTOSIS PROMOTING EFFECT

I. GENERAL REMARKS 151
II. RECENT OBSERVATIONS 151
III. MECHANISM OF INDUCTION OF LYMPHOCYTOSIS 154
    REFERENCES. 157

I. GENERAL REMARKS

One of the prominent features of whooping cough is a marked leukocytosis with a relative and absolute increase in lymphocytes. The first to record this lymphocytosis in pertussis was Fröhlich in 1897 (1). Since then, numerous workers have observed this phenomenon in children affected with whooping cough, as well as in experimental animals treated with B. pertussis (2,3). The peripheral leukocyte count in children can be greater than 175,000/mm^3 and the majority of cells are usually small lymphocytes (4). In animals, lymphocytosis has been induced by extracts from B. pertussis cells as well as with whole cell vaccines (5).

II. RECENT OBSERVATIONS

In recent years the lymphocytosis produced by pertussis vaccine has been reinvestigated by Morse and co-workers, and others. In his original work, Morse (6) employed whole cell pertussis vaccine con-

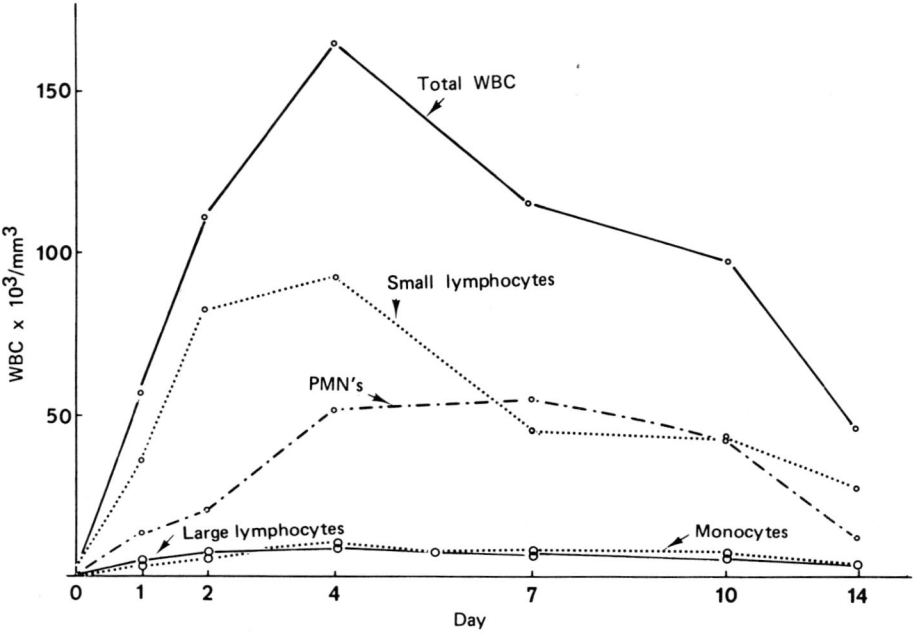

FIG. 1. Changes in the differential leukocyte counts of mice after the i.v. injection of 0.3 ml of pertussis vaccine. (Reprinted from Ref. 6, p. 53, by courtesy of The Rockefeller University Press.)

taining 50 to 100 x 10^9 cells/ml and male and female NCS mice from The Rockefeller University. This vaccine contained endotoxin and other biologically active substances that surely had some effects on the results obtained. The typical results after injection of vaccine are illustrated in Fig. 1. It is clearly seen that 0.3 ml of the vaccine stimulated 4 days after its administration a profound leukocytosis of over 150,000 cells/mm^3. Leukocytosis was detectable within 4 to 6 h after administration of the vaccine and dropped to 50,000 cells/mm^3 by the 14th day. As little as 2.5 x 10^9 cells (0.05 ml) of the vaccine given i.v. was effective. At the time of peak response, 4 days after injection, the small lymphocytes comprised 50 to 70% of the leukocyte population. Twenty-five to 40% were granulocytes, and the remainder were large lymphocytes and monocytes. Not all strains of mice respond as vigorously as the

II. RECENT OBSERVATIONS

NCS mice. We have generally observed a lower degree of leukocytosis in CFW and RML mice raised in our laboratory. Qualitatively, however, the response is similar (7).

It is interesting that the route of inoculation of the vaccines has a marked effect on the results observed (Fig. 2). The i.v. route is by far the most effective, the i.p. route is less effective, and the s.c. is the least effective (6). This relationship between route and efficacy is similarly observed in the sensitization of mice to histamine (8-10). Heating the vaccine at 100°C for 30 min completely destroyed its lymphocyte promoting activity while heating at 56°C for 30 min reduced activity only slightly (Fig. 3)(11). Our most highly purified preparations of pertussigen are active in submicrogram quantities in stimulating lymphocytosis. Various other workers have also made highly active preparations (11-15).

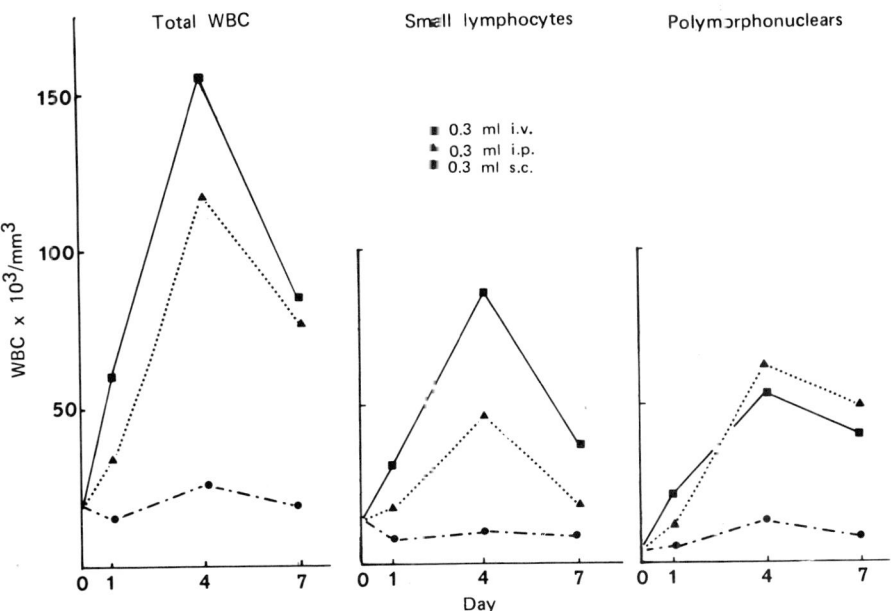

FIG. 2. Leukocyte counts of mice receiving 0.3 ml of pertussis vaccine by various routes. (Reprinted from Ref. 6, p. 56, by courtesy of The Rockefeller University Press.)

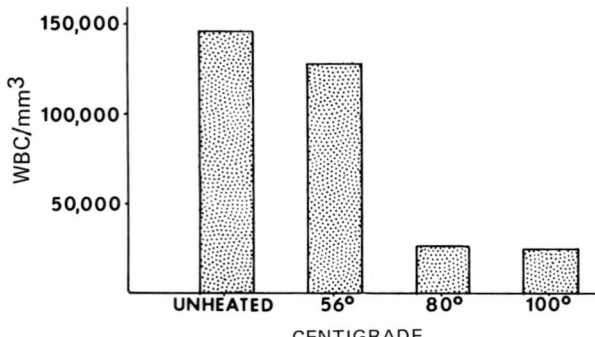

FIG. 3. Effect of heat on activity of LPF in *B. pertussis* culture supernatant fluid. The heat was applied for ½ h at the indicated temperature. (Taken from Ref. 11.)

III. MECHANISM OF INDUCTION OF LYMPHOCYTOSIS

The large number of circulating lymphocytes does not seem to come from the thoracic duct because its lymph flow is greatly decreased after injection of pertussis vaccine (16,17); however, the lymphocyte population of the thymus, lymph nodes, and spleen are depleted (16-18). A dramatic fall in thymus weight to approximately 35% of the controls was observed by the 4th day and it remained lower for over 14 days (Fig. 4). The mesenteric lymph nodes were also reduced in weight to about 57% of normal by the 7th day but by the 14th day they had returned to normal. The spleen, however, markedly increased in weight, and remained larger than spleens of control mice during the 14 days observed (6). Since endotoxin increases spleen size and weight, one wonders if this effect was mainly due to the endotoxin content of the vaccine. Histologically all 3 of the lymphoid organs studied -- thymus, lymph nodes, and spleen -- had disrupted follicular architecture and less lymphocytic elements. These changes were observed by the 4th day; by the 14th day a few normal-appearing follicles were seen in the lymph nodes. Splenic changes were also accompanied by a marked increase in red pulp, with numerous polymorphonuclear leukocytes in the sinusoidal spaces. This increase in red

III. MECHANISM OF INDUCTION OF LYMPHOCYTOSIS 155

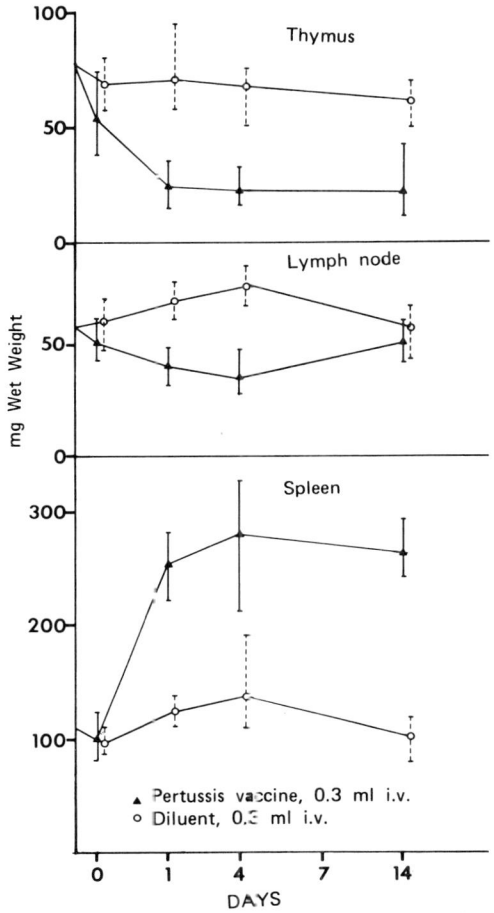

FIG. 4. Changes in weights of the thymus, mesenteric nodes, and spleens of mice receiving pertussis vaccine. (Reprinted from Ref. 6, p. 63, by courtesy of The Rockefeller University Press.)

pulp accounted for the increase in spleen weight. Because the number of mitotic cells in vaccinated mice did not differ from that in the controls, Morse believes that the increase in lymphocytes in the blood is due to depletion of lymphoid organs, rather than to a proliferation of cells. Iwasa et al. (16) also concluded that pertussigen induced the lymphocytosis by depleting the lymphatic tissues of their small lymphocytes and, because the thoracic duct drainage was almost completely blocked after administration of pertussigen, that these cells went directly from these organs to the blood. Lymphocytes treated in vivo or in vitro with their preparations did not migrate from the blood stream to the lymph nodes as did untreated lymphocytes. This phenomenon was studied by giving radioactive labeled lymphocytes i.v. and then recording the amount of radioactivity recovered in the thoracic duct cells. The change on the lymphocytes that makes them incapable to recirculate seems to be reversible, since after 2 weeks the lymph nodes are repopulated with lymphocytes (16,18,19). Neither prior thymectomy nor splenectomy reduced the leukocytosis induced by pertussis vaccine, but reticuloendothelial blockade reduced the intensity of the leukocytosis. This finding suggests that the uptake of *B. pertussis* cells by phagocytes is a necessary step (6). This observation should be repeated with soluble preparations of pertussigen because these may behave differently.

Mice receiving 2 s.c. doses of 10 to 20 x 10^9 cells given 1 week apart develop, 7 days after the 2nd injection, agglutinins to *B. pertussis* to a titer of 320 to 1,280. At this time, an i.v. injection of about 15 x 10^9 to 30 x 10^9 *B. pertussis* cells failed to elicit the leukocytosis observed in normal mice (11). Furthermore, anti-*B. pertussis* rabbit serum given i.v. 1 day before administration of pertussis vaccine also decreased the leukocytosis. These results would indicate that the LPF is antigenic, although direct proof for this was not given. This observation should be repeated with soluble preparations of pertussigen.

Neither x-ray treatment nor hydrocortisone, given to render the mice lymphocytopenic, prevented the leukocytosis induced by pertussis vaccine (20). However, x-irradiation of mice with leukocytosis induced by pertussis vaccine produced a prompt and extensive destruction of circulating as well as tissue small lymphocytes. These cells were cleared by the Kupffer cells of the liver (20).

Adler and Morse (21) have shown that erythrocytes and lymphocytes bind pertussigen and that this attachment is reversible. They suggest that reversible binding also occurs in vivo. This attachment may also occur with other particles such as starch granules, cellulose, and Sephadex (J. Munoz, unpublished observations).

Many have obtained semipurified preparations of pertussigen (5,11-15,22,23). In all cases the activities tested for (HSF, LPF, PA) were heat labile and partially destroyed by some proteolytic enzymes. Our purest preparations of pertussigen have been active in inducing leukocytosis and lymphocytosis in doses of 0.5 µg when given i.v.

REFERENCES

1. J. Fröhlich, *Jahrb., Kinderk., 54,* 53 (1897).
2. W. L. Bradford, H. W. Scherp, and M. R. Tinker, *Pediatrics, 18,* 64 (1956).
3. A. G. Ciplea, N. Pozsgi, Y. Faur, and V. Andreesco-Tigoiu, *Arch. Roum. Pathol. Exptl. Microbiol., 18,* 271 (1959).
4. J. H. Lapin, *Whooping Cough,* Charles C. Thomas, Springfield, Ill., 1943.
5. J. Munoz and R. K. Bergman, *Bacteriol. Rev., 32,* 103 (1968).
6. S. I. Morse, *J. Exptl. Med., 121,* 49 (1965).
7. C. R. Clausen, J. Munoz, and R. K. Bergman, *J. Bacteriol., 96,* 1484 (1968).
8. H. B. Maitland, R. Kohn, and A. D. MacDonald, *J. Hyg., 53,* 196 (1955).
9. J. Munoz, *Fed. Proc., 23,* 404 (1964).
10. J. Munoz and R. K. Bergman, *J. Immunol., 97,* 120 (1966).
11. S. I. Morse and K. E. Bray, *J. Exptl. Med., 129,* 523 (1969).
12. C. W. Parker and S. I. Morse, *J. Exptl. Med., 137,* 1078 (1973).
13. J. H. Morse and S. I. Morse, *J. Exptl. Med., 132,* 663 (1970).
14. Y. Sato and H. Arai, *Infect. Immun., 6,* 899 (1972).
15. S. Iwasa, S. Ishida, S. Asakawa, and M. Kurokawa, *Jap. J. Med. Sci. Biol., 21,* 363 (1968).
16. S. Iwasa, T. Yoshikawa, K. Fukumura, and M. Kurokawa, *Jap. J. Med. Sci. Biol., 23,* 47 (1970).
17. S. I. Morse and S. K. Riester, *J. Exptl. Med., 125,* 619 (1967).
18. S. I. Morse and B. A. Barron, *J. Exptl. Med., 132,* 663 (1970).
19. R. N. Taub, W. Rosett, A. Adler, and S. I. Morse, *J. Exptl. Med., 136,* 1581 (1972).
20. S. I. Morse, *J. Exptl. Med., 123,* 283 (1966).
21. A. Adler and S. I. Morse, *Infect. Immun., 7,* 461 (1973).

22. S. B. Lehrer, E. M. Tan, and J. H. Vaughan, *J. Immunol.*, *113*, 18 (1974).
23. M. Niwa, *J. Biochem. (Tokyo)*, *51*, 222 (1962).

Chapter 8

OTHER ACTIONS OF *BORDETELLA PERTUSSIS*

I.	INTRODUCTORY REMARKS	160
II.	HYPOGLYCEMIA	160
III.	HYPOPROTEINEMIA	162
IV.	INCREASED SUSCEPTIBILITY TO COLD STRESS	164
V.	EFFECT ON SUSCEPTIBILITY TO INFECTIONS	165
	A. Susceptibility to Bacterial Infections	165
	B. Nonspecific Protection to Bacterial Infections	166
	C. Effect on Viral Infections	166
	D. Adjuvant for Viral Vaccines	168
	E. Effect on Fungal Infections	169
VI.	EFFECT ON TUMORS	170
	A. Enhancement of Tumors	170
	B. Suppression of Tumors	171
VII.	INDUCTION OF INTERFERON	173
VIII.	INHIBITION OF MACROPHAGE RESPONSE TO BRAIN INJURY	175
IX.	HYPERSENSITIVITY REACTIONS TO *B. PERTUSSIS*	176
	A. Delayed Hypersensitivity	176
	B. Stimulation of Ascites	177
	C. Lung Edema	180
	REFERENCES	181

I. INTRODUCTORY REMARKS

This chapter discusses other effects of *B. pertussis* that have not been extensively investigated. In most cases the substance from the *B. pertussis* cell responsible for the effect is unknown; pertussigen may be responsible, but the role of other substances, especially endotoxin, has not been excluded.

II. HYPOGLYCEMIA

The finding of Parfentjev and Schleyer (1) that pertussis vaccine induces a pronounced hypoglycemia in mice has been confirmed by other workers (2,3). It has also been observed that *B. pertussis* cells induce a marked elevation of plasma insulin in rats and mice (4,5) which may be responsible for the hypoglycemia observed. The hypoglycemia was thought to be responsible for the sensitization of mice to histamine, serotonin, anaphylactoid, and anaphylactic shock (3), but we have been able to induce hypoglycemia with insulin (Table 1) without inducing histamine sensitization (6). Furthermore, Ganley (7) reported that alloxan diabetes protects pertussis vaccine-treated mice from histamine death, and attributed this protection to the elevation of blood sugar. We found, however, that a diabetogenic substance, d-mannoheptulose, that induces a marked hyperglycemia in pertussigen-treated mice does not protect mice from histamine death (Table 2). The induction of hypoglycemia has been demonstrated with purified preparations of pertussigen which most likely induces the hypoglycemia through its adrenergic blocking effects (8,9).

The onset and duration of hypoglycemia are illustrated in Fig. 1. As can be seen, within 2 h after administration of *B. pertussis* extract (BPE) a marked hypoglycemia can be detected that lasts for at least 21 days. The most pronounced histamine sensitivity follows closely the hypoglycemic state, although a hypersensitivity to histamine can be detected much later.

TABLE 1

Comparison of Histamine Sensitivity in Mice Made Hypoglycemic with Insulin or *B. pertussis* Extract (SE)[a,b]

Treatment	Depression of glucose level[c]	Histamine challenge (D/T)[d]
None	0	0/20
SE, 20 µg	-19	16/19
Insulin, 0.004 units	-22	0/10
Insulin, 0.008 "	-34	0/10
Insulin, 0.016 "	-78	0/10
Insulin, 0.04 "	-112	7/10
Insulin, 0.08 "	-119	7/9
Insulin, 0.16 "	-128	5/5

[a] Reprinted from Ref. 6, p 44, by courtesy of the Society for Experimental Biology and Medicine.

[b] SE was given i.v. 1 day before and insulin i.p. 30 min before obtaining blood samples or giving histamine (0.5 mg histamine base given i.p.).

[c] Differences between mean serum glucose values of experimental and normal control mice.

[d] Deaths/total mice tested.

TABLE 2

Persistence of Histamine Sensitivity in SE-Treated Mice Made Hyperglycemic with d-Mannoheptulose[a,b]

Treatment	Serum glucose compared to control	Histamine challenge (D/T)[c]
d-Mannoheptulose, 40 mg	+58	9/10
None	-19	9/10

[a] Reprinted from Ref. 6, p. 44, by courtesy of the Society for Experimental Biology and Medicine.

[b] Each mouse received i.v. 20 µg of SE 1 day before obtaining blood sample or challenging i.p. with 0.5 mg of histamine base. The treatment with d-mannoheptulose was given s.c. 26 h before obtaining blood sample or histamine challenge.

[c] Deaths/total mice tested.

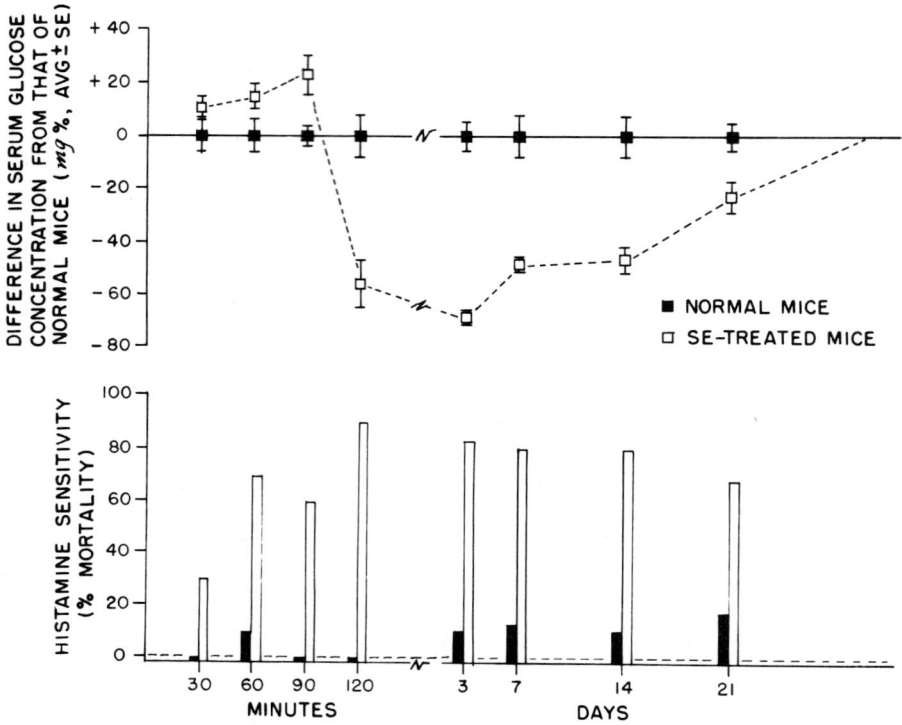

FIG. 1. Relationship between onset and duration of hypoglycemia and histamine hypersensitivity in saline extract (SE)-treated mice. The serum glucose values of the experimental mice are compared to those of normal mice set at a "zero" baseline. Normal mice had a serum glucose value of 184 ± 5.8 (avg. ± standard error) mg/100 ml serum during the experiment. (Reprinted from Ref. 6, p. 43, by courtesy of the Society for Experimental Biology and Medicine.)

III. HYPOPROTEINEMIA

Mice that receive B. pertussis extracts develop a marked reduction in plasma protein concentration (Fig. 2) that is mainly due to a decrease in albumin. Hypoproteinemia can be produced by purified

III. HYPROTEINEMIA   163

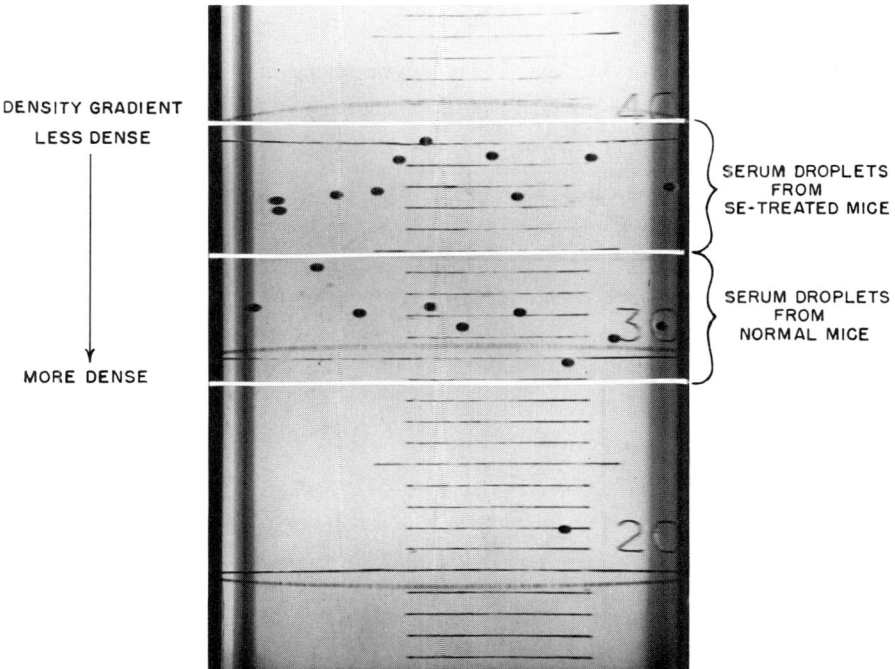

FIG. 2. Distribution of serum droplets in a density gradient showing distinct separation of serum droplets in the gradient dependent on whether they came from normal or from SE-treated mice. (Reprinted from Ref. 10, p. 965, by courtesy of the Society for Experimental Biology and Medicine.)

endotoxin preparations, and possibly this effect may be due to residual endotoxin in the extracts used. Purified pertussigen, however, also induces hypoproteinemia, but the significance of this effect is not well understood. Adrenal demedullated mice, which are fully sensitive to histamine, do not show hypoproteinemia (Table 3); thus, the hypoproteinemia may not be due to the same action of pertussigen that makes mice more susceptible to shock.

TABLE 3

Histamine Sensitivity and Total Serum Protein (TSP) Levels in Adrenal-Demedullated, Endotoxin-Treated, Pertussigen-Treated Mice[a]

Treatment	TSP[b] (g/100 ml serum)	Histamine sensitivity (D/T)[c]
Effect of adrenal demedullation		
"Sham" operated	6.7	1/10
Bilateral demedullation	6.6	8/10
Effect of endotoxin treatment		
Controls (no endotoxin)	6.7	0/10
Endotoxin, 10 μg, i.v.	5.6	1/10
Effect of pertussigen preparation		
Control	6.2	1/10
0.2 M fraction (10 HSD_{50s})	4.9	9/10

[a] Data from Ref. 10 and unpublished work.
[b] Average for 10 mice/group.
[c] Deaths/number of mice tested.

IV. INCREASED SUSCEPTIBILITY TO COLD STRESS

A number of years ago we described an increased susceptibility to cold induced in mice by pertussis vaccine (11). This was demonstrated by placing mice (10 mice per group) individually in tin cans and exposing them to 1 to 2°C. As can be seen in Table 4, mice that had been either adrenalectomized or pretreated with 2×10^9 cells of pertussis vaccine 4 days before submitting them to cold, died much sooner than control mice. Cortisone or hydrocortisone given i.p. 16 h before cold stress protected B. pertussis-treated and adrenalectomized mice. The steroids, however, did not protect pertussis-treated mice as well as adrenalectomized mice and failed to restore them to their normal resistance to cold. Although it has not been established whether pertussigen is responsible for this effect, it most likely is the active substance.

V. EFFECT ON SUSCEPTIBILITY TO INFECTIONS

TABLE 4

Survival of Adrenalectomized and *B. pertussis*-Treated Mice Exposed to Cold[a]

Group No.	Treatment[b]				
	I	II	III	IV	V
Adrenalectomized	-	+	-	-	-
Sham adrenalectomized	-	-	+	-	+
B. pertussis	-	-	-	+	+
Time of death at 1 to 2°C (h)	No. dead[c]				
1	0	0	0	0	0
2	0	0	0	1	2
3	0	2	1	7	5
4	0	2	0	2	3
5	0	3	1		
6	0	1	0		
7	3	2	1		
8	0		0		
>8	7		7		

[a]Reprinted from Ref. 11, p. 187, by courtesy of the Society for Experimental Biology and Medicine.
[b]+ = received treatment; - = did not receive treatment.
[c]Significance of difference: I vs. II, $P < 0.001$; I vs. III, not significant; I vs. IV, $P < 0.001$; II vs. IV, $P < 0.01$; IV vs. V, not significant.

V. EFFECT ON SUSCEPTIBILITY TO INFECTIONS

A. Susceptibility to Bacterial Infections

Pertussis vaccine given to mice renders them more susceptible to infections with certain Gram-negative bacteria that normally are nonpathogenic or only weakly so. Parfentjev and Arch (12,13), for example, found that pertussis vaccine-treated mice could be fatally infected with *Proteus vulgaris, Pasteurella multocida,* and *Pseudomonas fluorescens,* and Kind (14) produced fatal infections with *E. coli*. It is perhaps significant that all the infectious agents contain endotoxins to which mice become highly sensitive after receiving *B. pertussis* vaccine (14). It may well be that this hypersensitivity to endotoxin is responsible for the deaths caused by these relatively nonpathogenic bacteria.

B. Nonspecific Protection to Bacterial Infections

Landy (15) found that lipopolysaccharide extracted from *B. pertussis* increased the resistance of mice challenged 24 h later with *Salmonella typhosa* or with *B. pertussis* (16-18). Dubos and Schaedler (19) observed an increased susceptibility of mice to staphylococcus infection when treated a few hours before with *B. pertussis* vaccine, while, if the infection was given a few days later, an increased resistance was observed. It is interesting that Iida and Tajima (18) found that an i.p. administration of whole cell pertussis vaccine produced an increased resistance in mice against i.c. infection with *B. pertussis,* whereas i.p. injection of *B. pertussis* endotoxin did not induce any resistance to i.c. infection with *B. pertussis*. However, if the treatment with endotoxin was made i.c., resistance to infection developed. Both homologous and heterologous endotoxins, as well as double-stranded RNA complex of polyriboinosinic and polyribocytidylic acids (Poly I·C) given i.c. increased the resistance of mice to i.c. infection with *B. pertussis*. In the brains of animals thus treated, evident suppression of bacterial growth was seen.

C. Effect on Viral Infections

Parfentjev reported that pertussis vaccine also increased the susceptibility of mice to influenza virus (20) but nothing else has been reported along this subject.

Recently we found that administration of *B. pertussis* extracts (BPE) increased the incidence of abortive rabies in mice (21). In this work one group of mice received i.p. 40 µg of BPE mixed with

V. EFFECT ON SUSCEPTIBILITY TO INFECTIONS

virus and another group received only virus. Usually some mice, receiving virus only, fail to develop infection or immunity to subsequent i.c. challenge with rabies virus, whereas the majority develop clinical signs of infection. Although most of the latter group die, few recover and are subsequently immune to rabies. Figure 3 illustrates the results of one such experiment. BPE increased the recovery from rabies from 21% in the group receiving virus only to 71% in the group receiving BPE plus virus. BPE did not change the rate of noninfection.

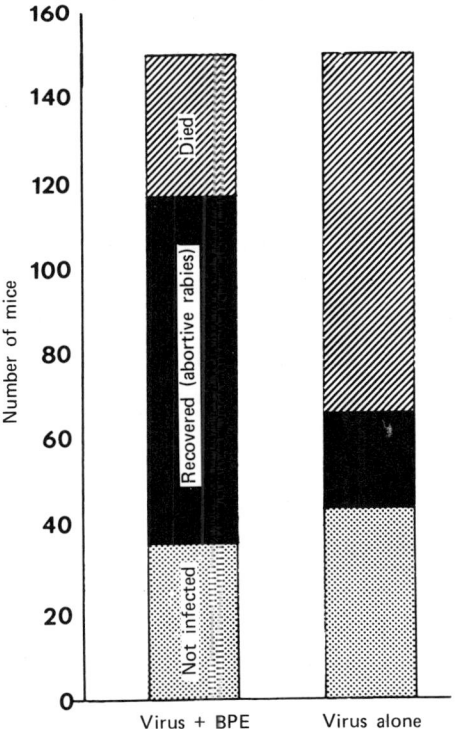

FIG. 3. Effect of B. pertussis extract (BPE) on the occurrence of abortive rabies in mice infected i.p. with street rabies virus (0.1 ml of a 10% suspension of infected mouse brain). 150 mice received virus plus 40 μg BPE and 150 mice received virus only. By abortive rabies is meant survival with various degrees of paralysis and subsequent immunity to reinfection with an i.c. challenge with 10^3 LD_{50} of fixed rabies virus, a challenge that invariably produces 100% deaths in normal mice. (Data from Ref. 21.)

TABLE 5

Effect of Single and Multiple Doses of B. pertussis Extract on the Occurrence of Abortive Rabies in Mice[a]

Group	BPE on day:				No. not infected[b]	Survivors/ infected	Abortive infection (%)
	-3	0	+3	+6			
1	+	-	-	-	13	5/37	13
2	-	+	-	-	4	31/46	67
3	-	-	+	-	10	8/40	20
4	+	+	+	-	6	7/44	16
5	-	+	+	+	2	28/48	58
6	-	-	-	-	12	5/38	13

[a]Reprinted from Ref. 21, p. 201, by courtesy of S. Karger AG, Basel.
[b]Fifty mice per group. Virus given on day 0.

BPE was effective in increasing the rate of abortive infection when given i.p., s.c., or i.v. It was effective only when given on the same day as the virus and not when given 3 days before or 3 days after virus (Table 5). As seen in Fig. 4, the amount of BPE required for this effect was about 8 µg per mouse.

D. Adjuvant for Viral Vaccines

A few workers have demonstrated that B. pertussis vaccine is a good adjuvant for virus vaccines. We (21) showed that BPE added to a killed rabies virus vaccine markedly improved its effectiveness to immunize mice against rabies (Fig. 5). The pertussis extract not only increased the protective potency of the killed vaccine but also increased the titer of anti-rabies virus antibody in the blood of mice receiving either killed or attenuated live virus vaccine.

V. EFFECT ON SUSCEPTIBILITY TO INFECTIONS

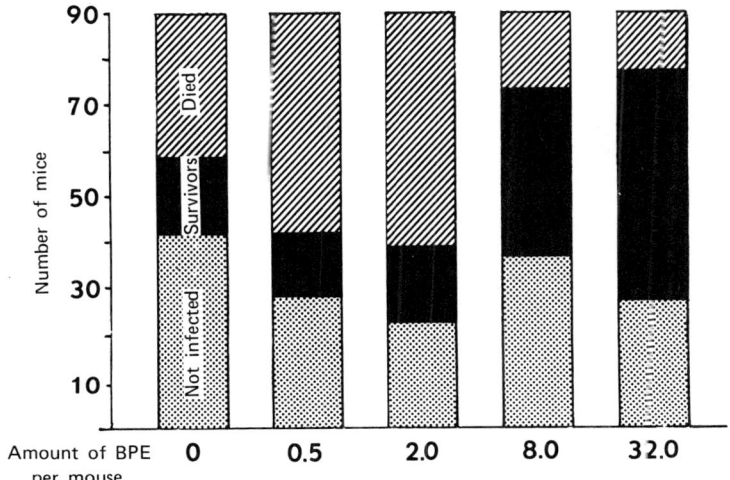

FIG. 4. Amount of B. pertussis extract effective in increasing abortive rabies in mice. This experiment was similar to that in Fig. 3. (Data from Ref. 21.)

Bogaerts and Durville-van der Oord (22) found that *B. pertussis* extracts enhanced the protective activity of encephalomyocarditis virus vaccine in mice as well as the antibody response to this vaccine.

E. Effect on Fungal Infections

Only few studies have been reported on the effect of *B. pertussis* vaccine on fungal infections. Abrahams (23) found that pertussis vaccine added to killed cryptococcus cells and, given to mice, greatly increased the immune response of these animals to subsequent infection with cryptococcus. Since the antibody response to the fungus was not enhanced, the authors concluded that resistance was primarily due to cellular immunity. Pertussis vaccine and extracts from *B. pertussis* cells also have induced a resistance to toxic challenge doses of *Candida albicans* (24).

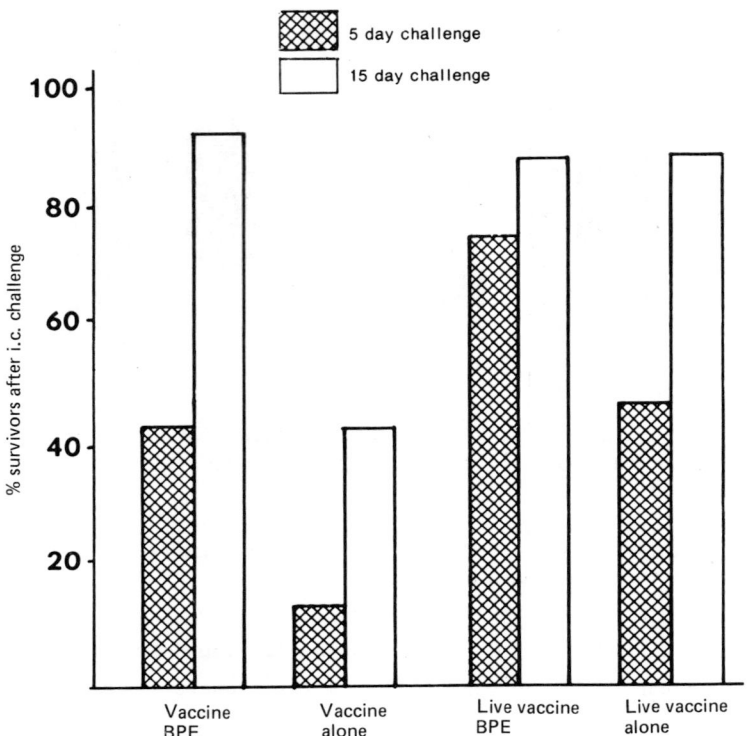

FIG. 5. Effect of *B. pertussis* extract on the effectiveness of killed and live rabies vaccines in mice. Groups of at least 40 mice were challenged 5 days or 15 days after vaccination. The challenge with virus was done by the highly effective intraplantar route. (Data from Ref. 21.) Live virus vaccine was chick embryo vaccine (Raboid, Fromm Labs., Inc., Grafton, Wisconsin). Killed virus vaccine was duck embryo vaccine (Eli Lilly and Co., Indianapolis, Indiana).

VI. EFFECT ON TUMORS

A. Enhancement of Tumors

B. pertussis vaccine has produced an increase in the growth of lymphomas in mice (25,26). A marked enhancement of growth of hepatoma in rats has also been observed when methylated albumin plus lethally irradiated tumor cells and pertussis vaccine were given 14 and 4 days before viable tumor cells were given s.c. (27). This

VI. EFFECT ON TUMORS

enhancement of tumor growth was not induced by the irradiated tumor cells alone or combined with methylated albumin; both pertussis vaccine and methylated albumin were essential. The pertussis vaccine alone or in combination with either irradiated tumor or methylated albumin was also ineffective. The effect was shown to be due to a humoral factor that could be absorbed with hepatoma cells, and thus, was most likely due to an enhancing antibody.

Similarly, Hirano et al. (28) found that pertussis vaccine given 4 days before lymphoma causes an augmented growth. It seems that this effect could have been due to stimulation of a tumor-enhancing antibody, but the authors point out that "pretreatment of animals with potent lymphoreticular stimulants causes 'pre-commitment' or 'syphoning off'...of immunologically competent cells...so that they are no longer available to react to subsequent antigenic challenge."

Using YLI lymphoma in inbred C57L/KL mice, Floersheim (25) found that the rate of growth and percentage of tumor takes was less in controls than in mice given 20×10^9 merthiolate-killed *B. pertussis* cells 30 to 60 min after inoculation with 10^4 to 10^6 tumor cells. Curiously, *B. pertussis* also increased the number of regressions, indicating a dual action of whole cell pertussis vaccine. The author seems to favor the idea that *B. pertussis* causes a suppression of the cellular mediated immune response.

B. Suppression of Tumors

Many reports indicating the usefulness of bacterial products on suppression of tumors in animals were published in the early 1930s and 1940s (29). This field of study has been renewed with the emphasis on finding a cancer "cure." Mycobacterial antigens as well as *Corynebacterium parvum* and killed *Coxiella burnetii* have been found effective in the rejection of syngenic hepatoma tumor (30,31, and D. Granger and E. Ribi, personal communication, 1975) in guinea pigs. The possibility that *B. pertussis* also has tumor-suppressing capacity was explored by Likhite (32). He used the mammary adenocarcinoma (line CAD_2, The Jackson Laboratory) originated in DBA/2 mice. When given s.c., 10^3 CAD_2 cells eventually kill all inoculated mice, while 10^6 to 10^7 cells kill recipients within 28 to 35 days. The tumor spreads readily and spontaneous regressions are never observed. Likhite found that when 26×10^9 killed *B. pertussis* cells were mixed with 10^7 tumor cells and given s.c. to mice, the tumors grew at normal rates during the first 14 days and then underwent rapid rejection and disappeared within 7 days. All the *B. pertussis*-treated mice survived, while 100% of the control animals receiving the tumor cells alone died within 28 to 35 days. The results obtained by the authors are illustrated in Fig. 6. The rapid and lasting

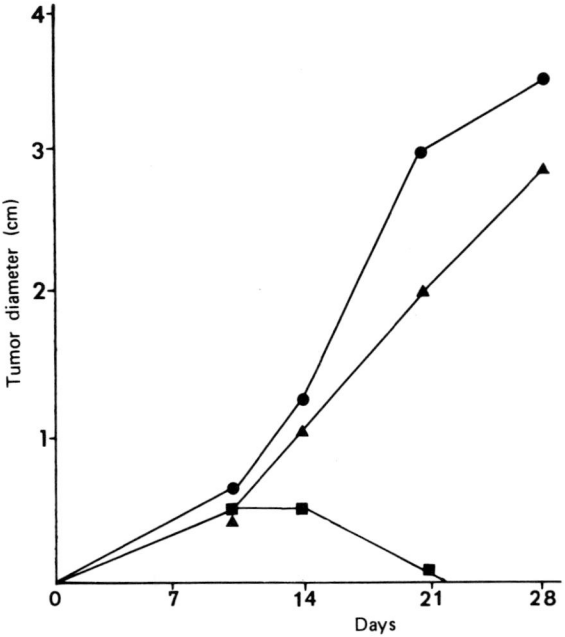

FIG. 6. Delayed rejection of s.c. transplants of tumor cells mixed with killed *B. pertussis* in DBA/2 mice. Each mouse received approximately 10^7 CAD_2 mammary adenocarcinoma cells and the following regimen: ■ = animals receiving 26×10^9 killed *B. pertussis* organisms mixed with the tumor cells; ▲ = animals treated with Hanks' solution; ● = animals treated with 26×10^9 killed *B. pertussis* cells given by i.p. route (this actually enhances tumor growth). (Taken from Ref. 32.)

rejection induced by B. pertussis was attributed to the immunological adjuvant effect of B. pertussis, but no differentiation was made between pertussigen and endotoxin. It is interesting that B. pertussis was effective only when given in close proximity with the tumor cells. According to Likhite, the close contact between tumor cells and B. pertussis antigens may have facilitated phagocytosis and the processing of tumor antigens by macrophages which are steps necessary for development of cellular and humoral immunity toward the mammary adenocarcinoma. It was shown that pertussis vaccine did not have any direct deleterious effect on the tumor cells and that a strong tumor-specific immunity to subsequent rechallenge developed in those mice rejecting the initial tumor. Likhite has also shown that B. pertussis induces rejection of other transplantable tumors in syngenic mice such as AKR lymphoma, L1210 acute lymphoblastic leukemia, and SAD sarcoma. B. pertussis also has induced rejection of rat tumors like the Shay chloroma, DMB 14 mammary adenocarcinoma, hepatoma, and 13762 mammary adenocarcinoma. Recently Yoo et al. (33) have also found that 2×10^9 cells given at the tumor site (Morris hepatoma 7777) produced a marked depression of tumor growth, but pertussis vaccine given at a distant site from the tumor caused an apparent enhancement of growth. This acceleration of tumor growth when pertussis vaccine is given at a site other than the tumor has been observed by others (25,34).

VII. INDUCTION OF INTERFERON

The induction of interferon by B. pertussis has also been reported (35,36). Kojima et al. (35) tested different purified components of B. pertussis cells for their ability to stimulate interferon in rabbits. A lipopolysaccharide extracted by the phenol-water method, a heat labile toxin preparation, and a protective antigen-HSF preparation induced interferon both in vivo and in vitro, while agglutinogen did not. It is interesting that heating preparations at 56°C for 30 min did not reduce the activity of the lipopolysaccharide or the protective antigen, but reduced the activity of HLT.

TABLE 6

Interferon Production by Purified Components of B. pertussis[a]

Preparation tested[b]	Pretreatment at 56°C for 30 min	Serum IF titer[c] in rabbits at 2 h	Fluid IF titer in cell culture of:		
			Spleen	Lymph node	Liver
OA	No	840	380	200	30
	Yes	800	nt	nt	nt
KA	No	<10	<10	<16	<10
			<10	<10	<10
HLT	No	340	630	1,900	80
			360	230	55
	Yes	43	67	nt	22
PA	No	1,200	930	240	50
	Yes	1,250	1,400	nt	50

[a] Data from Ref. 35.
[b] OA = O antigen; KA = K-agglutinogen; HLT = Heat labile toxin; PA = protective antigen; nt = not tested.
[c] IF titer was expressed as the reciprocal of the original dilution of the material showing 50% plaque reduction.

In vitro interferon was induced to a larger extent in spleen and lymph node cells than in liver cells (Table 6). The results seem to indicate that there are at least 2 substances in B. pertussis that can induce interferon: one a heat stable substance, probably endotoxin, and the other a highly heat labile substance which is probably the heat labile toxin. The preparations of PA were most likely contaminated with endotoxin and one cannot conclude from their data that PA was the substance responsible for inducing interferon.

VIII. INHIBITION OF MACROPHAGE RESPONSE TO BRAIN INJURY

Levine and Sowinski (37) showed that i.v. injection of pertussis vaccine inhibited the migration of macrophages into a zone of thermal coagulation and necrosis in the brain. The technique used to demonstrate this effect was as follows. A large thermal brain injury was produced through the intact exposed skull by applying with moderate pressure a preheated 37 ½ watt electric soldering iron for 7 sec to a rat under ether anesthesia. This treatment produced a large zone of coagulation necrosis in the cerebral cortex and subjacent structures. Three days later, normal rats had an abundant macrophage infiltration. Intravenous administration of pertussis vaccine 1 day before or immediately after burning the brain caused a marked decrease in the number of macrophages detected on the 3rd day. This effect was not due to endotoxin, but to a heat labile substance (destroyed at 80°C for ½ h). Levine used our semipurified preparations of pertussigen (HSF preparation in his paper) and found that they inhibited the macrophage migration to the necrotic area. The results with these fractions, prepared as described by us (38) are given in Table 7.

Levine and Sowinski also showed that i.v. administration of pertussigen produced a marked hyperplasia of the red pulp of the spleen with blurring of the normally clear demarcation from the perifollicular sheath. A depletion of small lymphocytes in the Malpighian corpuscles was observed and sometimes an increased extramedullary hematopoiesis and neutrophils in the red pulp. The splenic changes paralleled the inhibition of brain macrophages. The mechanisms of inhibition of migration of macrophages to the site of brain injury is not known.

TABLE 7

Inhibition of Rat Brain Macrophages by
Partially Purified Pertussigen[a]

Fraction	Dose per rat	Inhibition of macrophages
MgSO$_4$ precipitated HSF	400 µg	4/4[b]
	100 µg	2/4
	25 µg	0/4
1 M NaCl extract from MgSO$_4$ precipitate	400 µg	3/3
	100 µg	5/5
	25 µg	0/5
	6 µg	0/5
Whole cell vaccine[c]	0.1 ml	5/5
	0.025 ml	2/5
	0.006 ml	0/5

[a]Reprinted from Ref. 37, p. 352, by courtesy of the American Association of Pathologists and Bacteriologists and the American Society for Experimental Pathology.
[b]Inhibited/total tested.
[c]0.1 ml of vaccine contained approximately 0.4 mg of dry weight. All injections were given i.v. 1 day before thermal injury.

IX. HYPERSENSITIVITY REACTIONS TO B. PERTUSSIS

A. Delayed Hypersensitivity

In 1959 Rowley et al. (39) described a phenomenon of delayed hypersensitivity in the footpads and skin of rats receiving pertussis vaccine. Rats receiving 0.2 ml of pertussis vaccine (6 x 10^{10} cells per ml) equally divided into a front and hind foot, developed severe inflammation in 4 to 6 days. If these same rats are rechallenged in the other feet with 0.2 ml of vaccine containing 1.2 x 10^{10} cells per ml, a severe inflammation occurred within 24 h. In the 1st case,

IX. HYPERSENSITIVITY REACTIONS TO B. PERTUSSIS   177

the rats had become sensitized to the B. pertussis antigen before the reaction occurred, while in the 2nd case, the animals had already developed sensitivity and the delayed reaction appeared within 24 h. This type of sensitivity could be transferred with lymphoid cells but not with serum from sensitized donors and the inflammatory response was chiefly mononuclear. Pertussis vaccine could act as an adjuvant to induce sensitivity of a similar type to other antigens such as typhoid vaccine or gamma globulin. The reaction could not be suppressed with antiserotonin or antihistamine drugs but could be suppressed by x-irradiation.

We have recently observed similar reactions in the footpads and skin of mice receiving semipurified preparations of pertussigen. Five micrograms of our fraction AP (See Chapter 2) given s.c. in the footpad and 5 µg intradermally to normal mice induced an immediate inflammatory (edema) reaction that appeared within 4 h both in the foot and the skin site. This initial toxic effect disappeared within 48 h, but by the 5th day a marked inflammatory reaction developed at the site of the original injection of AP. This reaction termed "primary hypersensitivity response" by Rowley et al. gradually reduced in intensity but even 14 days after the initial injection the footpad and skin reactions could still be detected (Fig. 7). In presensitized mice that had received 10 µg of AP given i.p. 18 days before footpad testing, injection of 0.5 µg of AP into the right footpad (this test dose did not induce an initial toxic reaction) produced a marked delayed response at 24 h (Fig. 8). The marked swelling of the paws is illustrated in Fig. 9.

B. Stimulation of Ascites

The hypersensitivity response induced by pertussis vaccine seems to be responsible for the induction of marked ascites in rats observed by Levine and Gruenewald (40). These authors observed both the primary hypersensitivity response and the secondary response described by Rowley et al. (39). Rats receiving a single i.p. injection of 36×10^9 B. pertussis cells in 3 ml of saline showed accumulation

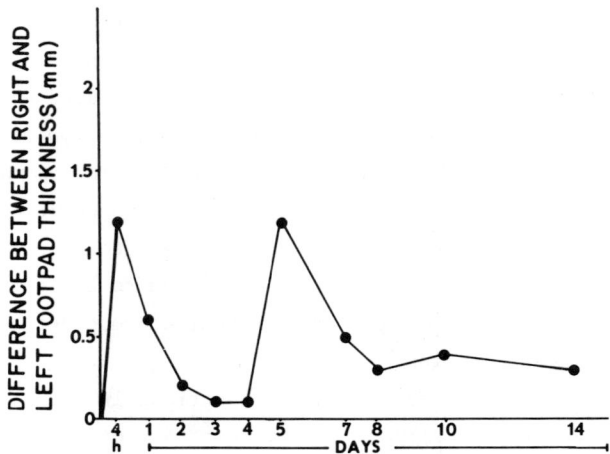

FIG. 7. Footpad tests in normal mice. Each point represents the average difference in thickness between left footpad receiving diluent only and the right which received AP. These mice received 5 µg AP in the right footpad and 5 µg intracutaneously on the back. (Data from J. J. Munoz and R. L. Cole, unpublished work.)

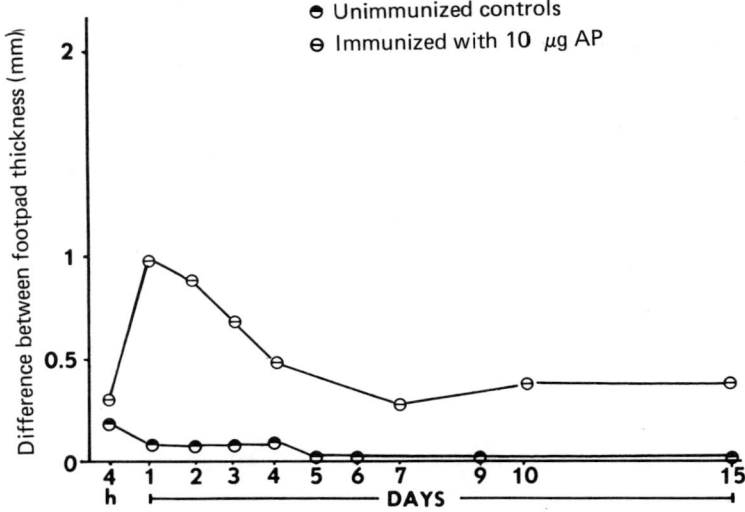

FIG. 8. Footpad tests in mice previosuly immunized with 10 µg AP given i.p. and tested 18 days later with 0.5 µg AP given in the right footpad. Each point represents the average difference in thickness between left (control) and right (experimental) footpads. (Data taken from J. J. Munoz and R. L. Cole, unpublished work.)

IX. HYPERSENSITIVITY REACTIONS TO *B. PERTUSSIS* 179

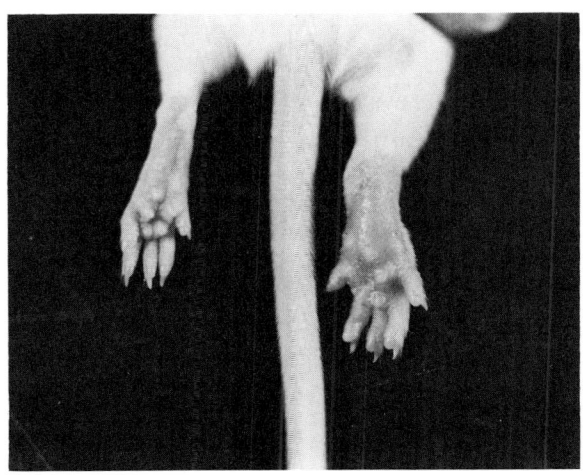

FIG. 9. Twenty-four hour reaction in the footpad of a mouse previously sensitized with AP, and tested with AP in the right and with buffered saline in the left footpad. (Taken from J. J. Munoz and R. L. Cole, unpublished work.)

of peritoneal fluid 5 to 8 days later. Rats tested by footpad inoculation with 0.1 ml of vaccine (6 x 10^9 cells) showed hypersensitivity reactions which appeared at the time when ascites was first detected. Histologic examination of the omentum 1 day after *B. pertussis* injection showed congestion, edema, infiltration by polymorphonuclear leukocytes, and lymphocytes. Those rats examined 2 to 13 days after inoculation exhibited more exudate and greater infiltration with lymphocytes, plasma cells, and histiocytes. By the 28th day the histological appearance of the omentum was normal. One day after injection of vaccine the mediastinal lymph nodes showed dilated sinuses with polymorphonuclear leukocytes and histiocytes swollen with ingested material. On succeeding days (2 and 4) an increased hyperplasia and the phagocytosis of nuclear fragments was observed and by the 5th day many sinuses were occluded by fibrin thrombi containing polymorphonuclear leukocytes and monoclear cells. By the 8th day the focal accumulation of histiocytes had the appearance of epithelioid cell granuloma. At 28 days there were many granulomas and some sinuses were filled with histiocytes and the nodes contained many plasma cells.

Rats sensitized 12 days prior to a 2nd identical i.p. dose of vaccine developed more intensive and earlier ascites. Most rats had from 6 to 24 ml of fluid by 4 to 6 days after the 2nd dose of *B. pertussis* cells. The histological changes were similar but more intense than those observed for the primary response.

C. Lung Edema

Edema in the lungs of mice receiving either live or killed *B. pertussis* cells or extracts has been observed by Andersen (41). She noticed that the lung weight after intranasal challenge with certain strains of *B. pertussis* increased as illustrated in Fig. 10. In this experiment maximum lung weight was achieved at 20 days, but from other experiments the optimal time was considered by Andersen to be 14 days after injection. With sterile extracts from *B. pertussis*

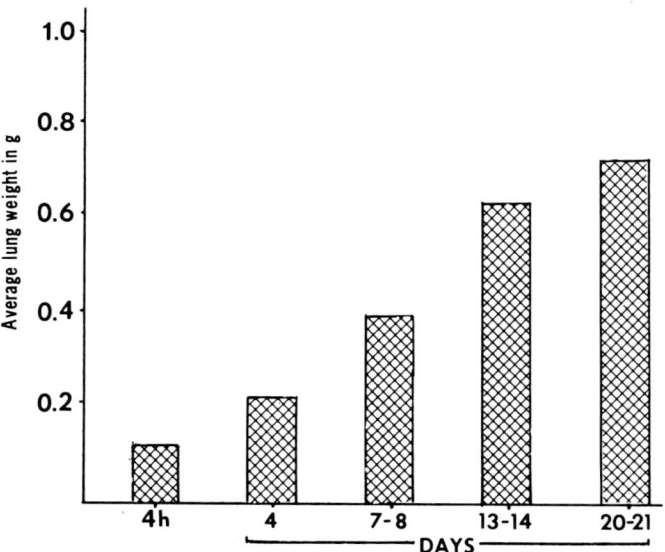

FIG. 10. Average lung weight at different times after intranasal infections with *B. pertussis*. Mice received intranasally approximately 14,000 cells from strain 5 tox and 5 mice sacrificed at each indicated time after infection (only 3 mice in the 20 to 21 day group). The lungs were removed and weighed. (Data from Ref. 41.)

she found the optimal time to be 7 days. Some *B. pertussis* strains were more effective than others, and 5-week-old mice were found to respond better than 8-week-old mice. A closely related bacterium, *B. parapertussis,* did not induce lung edema.

Although Andersen did not find an increased reactivity of the lung after previous immunization, this phenomenon seems to have a similar mechanism as the ascites production and hypersensitivity reactions previously described.

REFERENCES

1. I. A. Parfentjev and W. L. Schleyer, *Arch. Biochem. Biophys., 20,* 341 (1949).
2. M. G. Strong and M. Pittman, *J. Infect. Diseases, 96,* 152 (1955).
3. B. Gözsy and L. Káté, *Rev. Can. Biol., 23,* 427 (1954).
4. A. Gulbenkian, L. Schobert, C. Nixon, and I. I. A. Tabachnick, *Endocrinology, 83,* 885 (1968).
5. L. Muszbek, B. Csaba, and J. Csongor, *Acta Allergologica, 28,* 138 (1973).
6. R. K. Bergman and J. Munoz, *Proc. Soc. Exptl. Biol. Med., 131,* 42 (1969).
7. O. H. Ganley, *Can. J. Biochem. Physiol., 40,* 1179 (1962).
8. C. W. Fishel, A. Szentivanyi, and D. W. Talmage, in *Bacterial Endotoxins* (M. Landy and W. Braun, eds.), Quinn and Boden Co., Inc., Rahway, N.J., 1964, p. 474.
9. K. F. Keller and C. W. Fishel, *J. Bacteriol., 94,* 804 (1967).
10. R. K. Bergman and J. Munoz, *Proc. Soc. Exptl. Biol. Med., 131,* 964 (1969).
11. J. Munoz and L. F. Schuchardt, *Proc. Soc. Exptl. Biol. Med., 94,* 186 (1957).
12. I. A. Parfentjev, *6th Intern. Congress Microbiol.,* Rome, *3,* 38 (1953).
13. R. N. Arch and I. A. Parfentjev, *J. Infect. Diseases, 101,* 31 (1957).
14. L. S. Kind, *J. Immunol., 82,* 32 (1959).
15. M. Landy, *N. Y. Acad. Sci., 66,* 292 (1956).
16. E. K. Andersen, *Acta Pathol. Microbiol. Scand., 40,* 227 (1957).

17. O. G. Evans and F. T. Perkins, *Brit. J. Exptl. Pathol.*, *35*, 603 (1954).
18. T. Iida and M. Tajima, *Immunology*, *21*, 313 (1971).
19. R. J. Dubos and R. W. Schaedler, *J. Exptl. Med.*, *104*, 53 (1956).
20. I. A. Parfentjev, *Proc. Soc. Exptl. Biol. Med.*, *90*, 373 (1955).
21. J. F. Bell and J. J. Munoz, *Symp. Series Immunobiol. Standard.*, *21*, 199 (1974).
22. W. J. C. Bogaerts and B. J. Durville-van der Oord, *Infect. Immun.*, *6*, 513 (1972).
23. I. Abrahams, *J. Immunol.*, *96*, 525 (1966).
24. H. F. Hasenclever and E. J. Corley, *Sabouraudia*, *6*, 289 (1968).
25. G. L. Floersheim, *Nature*, *216*, 1235 (1967).
26. M. Hirano, J. G. Sinkovics, M. J. Ahearn, E. Shirato, and C. C. Schullenberger, *J. Reticuloendothel. Soc.*, *8*, 474 (1970).
27. R. A. Orlando, K. Craft, M. Glick, and R. W. Wissler, *Lab. Invest.*, *26*, 735 (1972).
28. M. Hirano, J. G. Sinkovics, C. C. Schullenberger, and G. D. Howe, *Science*, *158*, 1061 (1967).
29. H. C. Reilly, *Cancer Res.*, *13*, 821 (1953).
30. B. N. Halpern, G. Biozzi, C. Stiffel, and D. Mouton, *Nature*, *212*, 863 (1966).
31. B. Zbar, I. Bernstein, T. Tanaka, and H. J. Rapp, *Science*, *170*, 1217 (1970).
32. V. V. Likhite, *Cancer Res.*, *34*, 1027 (1974).
33. T. J. Yoo, C. Y. Kuo, and J. I. Stagner, *Lancet*, *1*, 402 (1975).
34. S. Malkiel and B. J. Hargis, *Cancer Res.*, *21*, 1461 (1961).
35. Y. Kojima, F. Yoshida, and Y. Nakase, *Jap. J. Microbiol.*, *17*, 160 (1973).
36. L. Borecký and V. Lackovič, *Acta Virol.*, *10*, 271 (1966).
37. S. Levine and R. Sowinski, *Amer. J. Pathol.*, *67*, 349 (1972).
38. J. J. Munoz, R. F. Smith, and R. L. Cole, *Symp. Series Immunobiol. Standard.*, Vol. 13, Karger, Basel, 1970, p. 265.
39. D. A. Rowley, J. Chutkow, and C. Attig, *J. Exptl. Med.*, *110*, 751 (1959).
40. S. Levine and R. Gruenewald, *Exptl. Mol. Pathol.*, *1*, 104 (1962).
41. E. K. Andersen, *Acta Pathol. Microbiol. Scand.*, *40*, 248 (1957).

Chapter 9

HYPOTHESIS ABOUT THE MODE OF ACTION OF PERTUSSIGEN
AND FUTURE WORK ON SUBSTANCES FROM *BORDETELLA PERTUSSIS*

I. HYPOTHESES ON ACTIONS OF PERTUSSIGEN 183
    A. Introductory Remarks. 183
    B. Hypersensitivity to Vasoactive Amines and Shock 184
    C. Enhancement of Autoimmune Diseases. 185
    D. Adjuvant Action of Pertussigen. 185
    E. Increased Susceptibility to Anaphylaxis 186
    F. Mouse Protection. 186
    G. Other Effects 187
II. FUTURE WORK. 188
    REFERENCES 189

I. HYPOTHESES ON ACTIONS OF PERTUSSIGEN

A. Introductory Remarks

Most of the work done on the mechanisms of action of pertussigen has been done with whole cell vaccines. The complexity of these vaccines has made the interpretation of results difficult. Even in studies employing semipurified pertussigen, the role of contaminating substances, such as endotoxin, has influenced the results. A reasonably clear picture of the mode of action of pertussigen can be obtained, however, by segregating the effects of heat stable substances (mainly endotoxin) from those produced by heat labile (80°C for ½ h) substances (mainly pertussigen, since the HLT can easily be eliminated). In this manner, and with the analysis of the few studies done with purified pertussigen, it has been possible to establish that pertussigen is responsible for many of the biological effects of *B. pertussis* cells.

All the various effects which pertussigen produces in mice and rats may be manifestations of a single fundamental derangement in the animal body. However, this is difficult to support on the basis of current information. We believe that at most 2 main functions are responsible for the actions of pertussigen. One is the increase in antibody production (adjuvant action) and the other an interference with the action of epinephrine. Let us analyze the most important phenomena induced by pertussigen in light of these 2 major effects.

B. Hypersensitivity to Vasoactive Amines and Shock

The increased susceptibility to vasoactive amines definitely seems to be due to an interference of pertussigen with the function of epinephrine. Pertussigen does not interfere with the production of the hormone, but with its action. Pertussigen accomplishes this apparently by binding to the adrenergic receptors of the cells and preventing them from responding normally to epinephrine. According to Fishel (1), the blocking occurs at the β-receptors, although there is no consensus as to whether the β-receptors are exclusively involved (2,3). There is, however, no doubt that epinephrine function is involved (4-6).

An important consequence of this blocking effect seems to be the increased vascular permeability observed in the skeletal muscle of mice treated with pertussigen (7). But perhaps the most important change as a result of the blockade of epinephrine action is a decreased ability to counteract the vascular permeability induced by vasoactive amines (7). This inability to compensate for the changes induced by vasoactive amines must be responsible for the increased susceptibility to shock. This shock can be induced by exogenous administration or endogenous release of the vasoactive substances.

I. HYPOTHESES ON ACTIONS OF PERTUSSIGEN

C. Enhancement of Autoimmune Diseases

The enhancement of EAE must involve the 2 main actions of pertussigen: the adjuvant action to encephalitogenic antigen and the increased susceptibility to the action of vasoactive amines due to an adrenergic blockade.

We have shown that when symptoms of EAE are seen in pertussigen-treated rats a marked increase in the vascular permeability in the spinal cord occurs (8). At this time one can also demonstrate an increased concentration of albumin and the presence of globulins in the cord itself. The increased production of antibodies, found either free or on sensitized lymphoid cells, in conjunction with the increased vascular permeability can, it seems to us, explain the increased susceptibility to EAE. Since in the hyperacute form of EAE complete recovery of fully paralyzed rats can occur, permanent damage to the nervous system does not seem to occur, and paralysis is probably due only to an interference with the transmission of nerve impulses to the muscle. When this "short circuiting" is corrected, nerve and muscle function are restored and the paralysis disappears.

D. Adjuvant Action of Pertussigen

We consider the adjuvant action of pertussigen to be responsible for many of the fundamental changes produced by this substance. It is, however, difficult with the evidence at hand to understand how pertussigen acts as an immunological adjuvant. We have suspected that the increase in vascular permeability may make the antigens more effective by allowing them to reach the antibody-forming cells more quickly and in greater concentration before being eliminated from the body. If this were the case, the only fundamental action that pertussigen would need to possess is the blocking of epinephrine action. However, no conclusive experimental evidence to show that epinephrine blockade increases antibody formation is yet available, although some have reported that adrenalectomy in rabbits increases antibody formation (9).

The adjuvant action could also be due to an increase in the population of antigen responsive cells (10-12) or to an actual combination of pertussigen with the antigen to make it more effective. Little is known about the action of pertussigen on these mechanisms and they should be investigated.

The accelerated enhancement of IgE is another action of pertussigen which is at present not understood.

E. Increased Susceptibility to Anaphylaxis

Increased susceptibility to either passive or active anaphylaxis is undoubtedly explained on grounds similar to those for enhanced susceptibility to shock. However, in active anaphylaxis the stimulation of antibody response is an equally important action.

F. Mouse Protection

The mouse protective effect of pertussigen is also of great interest to us. This substance in microgram doses induces "active" immunity to i.c. challenge with virulent *B. pertussis*. We think that it is possible that this protection is due to a combination of 2 effects: a nonspecific effect that could be an increased vascular permeability of the blood-brain barrier, and a specific effect due to an increase in antibody production against surface antigens of the *B. pertussis* cell. Mice are not normally susceptible to infection with *B. pertussis* cells given by i.p., i.v., or s.c. routes. Large numbers of cells given intranasally do kill mice, while small challenges do not, in spite of multiplication of the bacteria (13,14). Resistance to intranasal infection can be induced by prior specific immunization but the antigen responsible (heat resistant) is different than the heat labile substance responsible for the induction of protection to i.c. challenge (15). The nonspecific effect of pertussigen may play a very important role in the protection against the i.c. challenge by increasing the amount of antibody and leukocytes capable to reach the brain to clear the infection. It is interesting to mention here the observations made by Iida and Tajima (16) who found that endotoxins from B. pertussis, S. typhimurium, or B. bronchiseptica which are totally unable to induce immunity in mice when given i.p. were able to protect mice when given i.c. 48 h before infection by the same route. Even the polyriboside, Poly I•C, which has activities similar to endotoxin, was capable of protecting mice under similar conditions. These observations, if confirmed, may indicate that changes in the brain that allow normal defense mechanisms to reach it (phagocytic cells and antibodies) may nonspecifically protect against B. pertussis.

Pertussigen then could induce protection specifically by increasing antibody production to surface antigens of B. pertussis and nonspecifically by increasing the permeability of the blood-brain barrier (17,18). In the absence of pertussigen other antigens of B. pertussis appear to be rather ineffective in inducing active immunization to i.c. infection.

G. Other Effects

The mechanism of lymphocytosis induced by pertussigen is not well understood. It is thought that the lymphocytes come only from preformed lymphocytes in the thymus, lymph nodes, and spleen and that these lymphocytes in the presence of pertussigen can not return to the lymph organs (19,20). Pertussigen adsorbs to lymphocytes as well as red cells and other particles (21) and whether this has an effect on their inability to recirculate is still questionable. We believe instead that the permeability changes induced by pertussigen as a result of the blocking action on epinephrine are somehow responsible for lymphocytosis, but this is still highly speculative.

The hypoglycemia induced by pertussigen must be due to its adrenergic blocking effect, since failure of epinephrine to act would induce changes in sugar metabolism (22).

The induction of hypoproteinemia may be due mainly to endotoxin, because endotoxin or heated B. pertussis extracts induce this response. The manner in which this happens is not at all clear at this point.

II. FUTURE WORK

From what has been said in this book, it is clear that many aspects of the *B. pertussis* field are still poorly known. Many problems have remained partially solved and many others are for the most part totally unexplored. One substance that should be studied further is the heat labile (56°C for ½ h) toxin. Little is known about its exact chemical composition, its mode of action, its effect on lymphoid tissue and antibody production, its production and distribution among *Bordetella* species, etc. The marked toxic action on the spleen and possibly lymph nodes makes the HLT an interesting material for future immunological studies.

Other substances that require further work are the agglutinogens. Their role on induction of active immunity to whooping cough, their toxic activity, their antigenicity, and their chemical identity are not known.

Hemagglutinin also should be further investigated, especially since recently it has been reported that this factor may be identical to pertussigen (23). Thus, the relationship of pertussigen to this substance should be thoroughly investigated and if found to be different, its biological activities and chemical nature studied. The endotoxin from *B. pertussis* has only been poorly investigated and much work is needed on this substance.

Many aspects of pertussigen should be investigated more thoroughly. Improvement of the methods of purification to increase yields are needed. Its complete characterization, its exact mode of action, and its role in immunity to whooping cough should be further studied. It is also important to establish whether pertussigen is identical in all strains of *B. pertussis* or if differences occur which make it a family of compounds with slightly different chemical composition but similar biological activities.

The many biological activities of pertussigen have opened new and promising research horizons. For example, its epinephrine-blocking effects make pertussigen a useful tool to study the role of epinephrine in various conditions in mice. It may actually be useful as a drug in the treatment of certain hypertensive conditions, since in mice at least its blocking action can be produced by microgram quantities and the blockade lasts for many weeks. The adjuvant effect may make it useful as an adjuvant for various vaccines, and its ability to stimulate IgE is an important property in studies of allergic problems.

A better understanding of the fundamental action of pertussigen may give some insight on mechanisms of production and stimulation of IgE, on mechanisms of hyperacute EAE and other autoimmune diseases, on mechanisms of induced lymphocytosis, and on methods for prevention and treatment of shock.

Along these lines, it has already been proposed that the B. pertussis-treated mouse can be used as an animal model for the study of asthma (22).

Other interesting effects of B. pertussis which should be studied further are the enhancement and suppression of tumor growth and hypersensitivity reactions.

As can be seen, the B. pertussis problem touches many disciplines and thus makes it important to many facets of the medical field. We hope that this book will stimualte further research on these interesting problems.

REFERENCES

1. C. W. Fishel, A. Szentivanyi, and D. W. Talmage, *Bacterial Endotoxins* (M. Landy and W. Braun, eds.), Rutgers University Press, New Brunswick, N.J., 1964, p. 474.
2. R. K. Bergman and J. Munoz, *Life Sci.*, *10*, 561 (1971).
3. C. W. Parker and S. I. Morse, *J. Exptl. Med.*, *137*, 1078 (1973).
4. J. Munoz and L. F. Schuchardt, *J. Allergy*, *25*, 125 (1954).
5. R. K. Bergman and J. Munoz, *Nature*, *205*, 910 (1965).
6. R. K. Bergman and J. Munoz, *Proc. Soc. Exptl. Biol. Med.*, *122*, 428 (1966).
7. R. K. Bergman and J. Munoz, *J. Allergy Clin. Immunol.*, *55*, 378 (1975).
8. R. K. Bergman and J. Munoz, *IRCS Med. Sci.*, *4*, 202 (1976).

9. D. F. B. Char and V. C. Kelley, *Proc. Soc. Exptl. Biol. Med.,* *109,* 599 (1962).
10. H. Finger, P. Emmerling, and E. Brüss, *Infect. Immun., 1,* 251 (1970).
11. D. W. Dresser and J. M. Phillips, in *Immunopotentiation,* Ciba Found. Symp. 18 (New Series, Elsevier, 1973), pp. 3-28.
12. C. E. Reed, M. Benner, S. D. Lockey, T. Euta, S. Makino, and R. H. Carr, *J. Allergy Clin. Immunol., 49,* 174 (1972).
13. J. M. Dolby, D. W. Thow, and A. F. B. Standfast, *J. Hyg. Camb., 59,* 191 (1961).
14. M. Pittman, in *Infectious Agents and Host Reactions* (Mudd, ed.), W. B. Saunders Co., Philadelphia, 1970, pp. 239-270.
15. A. F. B. Standfast, *Immunology, 2,* 135 (1958).
16. T. Iida and M. Tajima, *Immunology, 21,* 313 (1971).
17. M. C. Berenbaum, J. Ungar, and W. K. Stevens, *J. Gen. Microbiol., 22,* 313 (1960).
18. T. Iida, N. Kusano, A. Yamamoto, and M. Konosu, *J. Pathol. Bacteriol., 92,* 359 (1966).
19. S. I. Morse and B. A. Barron, *J. Exptl. Med., 132,* 663 (1970).
20. S. Iwasa, T. Yoshikawa, K. Fukumura, and M. Kurokawa, *Jap. J. Med. Sci. Biol., 23,* 47 (1970).
21. A. Adler and S. I. Morse, *Infect. Immun., 7,* 461 (1973).
22. A. Szentivanyi, *J. Allergy, 42,* 203 (1968).
23. Y. Sato, H. Arai, and K. Suzuki, *Infect. Immun., 7,* 992 (1973).

Appendix

MATERIALS AND METHODS USED IN OUR WORK

I.	MEDIA FOR CULTIVATION OF *B. PERTUSSIS*	193
	A. Solid Medium: B-G Agar.	193
	B. Liquid Medium.	193
	C. Diluent for Challenge Suspensions.	194
II.	PREPARATION OF CELL SUSPENSIONS	194
III.	TESTS FOR BIOLOGICAL ACTIVITY	195
	A. Mouse Protection	195
	B. Increased Sensitivity to Histamine	196
	C. Increased Sensitivity to Serotonin	197
	D. Increased Sensitivity to Serotonin-Histamine	197
	E. Leukocytosis-Promoting Activity.	197
	F. Hypoproteinemia.	198
	G. Hypoglycemia	198
IV.	METHODS TO MEASURE VASCULAR PERMEABILITY.	198
	A. Evans Blue Dye Method.	198
	B. Radiolabeled Human Serum Albumin (HSA) Method.	199
V.	DIRECT AGGLUTINATION OF ERYTHROCYTES.	200
VI.	ANTIBODY TITRATION BY THE BIS-DIAZO-BENZIDINE (BDB) TECHNIQUE	200
	A. Preparation of BDB	200
	B. Conjugation of Erythrocytes to Antigen	201
	C. Preparation of Buffers	201
	D. Hemagglutination Test.	201

VII.	DETECTION OF ANTIBODY-FORMING CELLS IN SPLEEN AND LYMPH NODES 202
	A. General Remarks. 202
	B. Preparation of Chicken Anti-SRBC Gamma Globulin. . . 202
	C. Guinea Pig Complement. 202
	D. Sheep Erythrocytes 203
	E. Balanced Salt Solution (BSS) 203
	F. Amplifying Serum 203
	G. Collection and Preparation of Lymphocytes. 204
	H. Performance of the Test. 204
VIII.	PASSIVE CUTANEOUS ANAPHYLAXIS (PCA) IN MICE 205
IX.	ANAPHYLAXIS IN *B. PERTUSSIS*-TREATED MICE. 206
	A. Active Anaphylaxis 206
	B. Passive Anaphylaxis. 206
X.	PREPARATION OF ANTISERA 206
	A. Antisera to Purified Proteins. 206
	B. Antisera to *B. pertussis* 207
XI.	AGGLUTINATION TESTS 207
XII.	AGGLUTININ PRODUCTION IN MICE 208
XIII.	AGGLUTINOGEN ABSORPTION TEST. 208
XIV.	IMMUNIZATION OF MICE TO PRODUCE ASCITIC FLUID CONTAINING IgE ANTIBODIES 209
XV.	ENHANCEMENT OF EXPERIMENTAL ALLERGIC ENCEPHALOMYELITIS. 210
XVI.	SCHULTZ-DALE REACTION 210
XVII.	GEL DIFFUSION TEST. 211
XVIII.	IMMUNOELECTROPHORESIS (IEP) 211
XIX.	DISC ELECTROPHORESIS IN ACRYLAMIDE GEL. 212
XX.	HYDROXYLAPATITE COLUMN CHROMATOGRAPHY 215
XXI.	STARCH BLOCK ELECTROPHORESIS. 215
XXII.	ZONAL DENSITY GRADIENT ELECTROPHORESIS. 216
	REFERENCES. 216

I. MEDIA FOR CULTIVATION OF B. PERTUSSIS

A. Solid Medium: B-G Agar

We make this medium as follows: to 1 liter of distilled water are added 30 g of B-G agar base (Difco), 10 g of neopeptone (Difco), and 10 ml of glycerol. The solution is sterilized at 15 lb pressure for 20 min, cooled to about 45 to 50°C, and 150 ml of fresh (not over 1 week old) defibrinated horse blood are added. It is mixed well and poured in 10- to 15-ml aliquots per standard size Petri dish. Slants or Blake bottles can also be used.

B. Liquid Medium

Casamino acid medium is a modification of Cohen-Wheeler (1) medium and is made as follows:

Bacto casamino acid (Difco technical)	14 g
Bacto yeast extract (Bacto)	3 g
Soluble starch (indicator grade from Difco)	1.0 g
Niacin	0.02 g
Glutathione	0.01 g
$MgCl_2 \cdot 6H_2O$	0.1 g
$CaCl_2 \cdot 2H_2O$, 1% solution	1 ml
$FeSO_4 \cdot 7H_2O$, 0.05% solution	2 ml
$CuSO_4 \cdot 5H_2O$, 0.05% solution	1 ml
Trizma base	6 g
Distilled water added to bring volume to	1 liter

The pH is adjusted to 7.5 with 2 N HCl and the medium heated until the starch is dissolved. Dispense 200 ml into 1 liter Blake bottles or 650 ml into 2-liter Erlenmeyer flasks. Stopper with gauze-covered cotton stoppers and autoclave at 15 lb for 25 min.

It should be stressed that not all lots of casamino acids or yeast extracts are satisfactory and each lot is tested for its ability to support growth.

C. Diluent for Challenge Suspensions

Diluent for making challenge suspension is made as follows: Three grams of casamino acid (Difco certified) are dissolved in 100 ml of distilled water and the pH adjusted to 7.0 to 7.2 with NaOH. The medium is dispensed in 100-ml amounts in 250-ml Erlenmeyer flasks and autoclaved at 15 lb for 20 min.

II. PREPARATION OF CELL SUSPENSIONS

A smooth strain of *B. pertussis* is used. We keep all our cultures in a lyophilized form under vacuum at -15°C. A vial of the lyophilized culture is reactivated by adding about 0.1 ml of sterile water or 3% casamino acid solution and resuspended. Two to 3 drops of this suspension are transferred to a B-G agar plate, spread with a sterile wire loop, and the plates incubated for 3 days at 37°C. The plates are kept in a plastic bag to prevent dehydration and external contamination. On the 3rd day, the culture is transferred to another B-G plate and incubated for 2 days at 37°C and then this culture is used to prepare inocula for either solid or liquid media. To grow cells in solid media a suspension of cells is made in casamino acid medium and 0.2 ml of this is added to each Petri plate containing B-G agar and spread with a glass spreader made from a Pasteur pipet. If large Blake bottles are used, about 2 ml of inoculum per flask is used and the inoculum spread by tilting the bottle back and forth. In agar we prefer to incubate cells for not longer than 48 h at 37°C. At this time the cells are collected by adding 3 to 5 ml of saline and removing the growth by means of a glass capillary bent at right angles. Cells grown in Blake bottles can be removed by means of saline and glass beads added to the bottle. The collected cells are filtered through 4 layers of gauze, centrifuged and resuspended in

the desired liquid, and used either in this form or killed by adding
1/10,000 merthiolate. To make suspension for agglutination tests we
prefer to suspend the cells in 0.3% formalin to prevent autolysis.

Liquid medium is preferable for growing large amounts of cells.
Inoculum for liquid medium is made by transferring B-G agar-grown
cells to a 1 liter Blake bottle (heavy inoculum is needed) containing
200 ml of casamino acid medium. The flask is incubated on its widest
side with shaking in a reciprocal shaker for 2 to 3 days. When good
growth is obtained, this material can be used to inoculate as many
1 liter Blake bottles or 2-liter Erlenmeyer flasks as needed. About
10 ml per Blake bottle or 20 ml per Erlenmeyer flask are needed.
Incubation is carried out at 35 to 37°C with constant shaking for
not longer than 3 days. The cells are collected by centrifugation.
The cell paste can be kept frozen or the cells can be dialyzed and
lyophilized. In most of our work we have treated the cells with
acetone and air dried them (see Section 2-VII). In this form the
biological activities of these cells remain stable for many years
when kept in a dry atmosphere at room temperature.

III. TESTS FOR BIOLOGICAL ACTIVITY

A. Mouse Protection

This test follows with minor changes the test devised by Pittman
and Lieberman (2).

We employ 21-day-old mice of either RML or CFW strains raised
in our laboratory. The mice are preferably one sex but in many cases
we have used equal numbers of both sexes. The mice are randomized in
glass jars adequate in size for 5 to 6 mice. These jars contain
corncob bedding and the mice are allowed Purina chow pellets and
water ad lib. Three dilutions of the material to be tested are used
and 16 mice per each dilution are immunized i.p. with 0.2 ml of the
dilution. Two weeks later the mice are challenged i.c. with approximately 30,000 colony-forming units of a *B. pertussis* culture (a
substrain of culture 18323 originally obtained from Dr. P. Kendrick).

The mice are observed for 14 days. Deaths during the first 3 days are considered to be from trauma and those occurring later result from specific infection. After 14 days the survivors are recorded and the PD_{50} calculated. Mice that show signs of infection on day 14 are counted as dead. A group of 16 unimmunized mice is also challenged with the same dose of *B. pertussis* to serve as virulence control.

The preparation of the cell suspension for i.c. challenge is very important and is done as follows: A lyophilized culture is revived as described above. The 1st B-G plate is incubated for 3 days. A transfer is made to a 2nd B-G plate and incubated for 2 days. From this plate a 3rd plate is heavily inoculated and incubated for 24 h. The growth on this plate is used for challenge by scraping the growth carefully from the surface of the plate and suspending it in 3% casamino acid. The suspension is standardized with a Coleman Jr. spectrometer at a wavelength of 550 nm. From a standard curve, made by plotting optical density against numbers of cells in known suspension (made by direct count) of *B. pertussis* cells, the number of cells per milliliter are calculated and appropriate dilutions are made to obtain a challenge suspension containing 10^6 cells per ml. Three-hundreths milliliter of this suspension is used to challenge each mouse (approximately 30,000 colony-forming units). The viable count is checked by preparing a 1/2,500 dilution of the challenge suspension and plating out 0.1 ml on each of 3 B-G agar plates. The count per 0.1 ml should be 40 colonies per plate, but actually the counts range from 30 to 50. All dilutions are made in 3% casamino acid diluent. The challenge suspension is kept cold and used within 2 ½ h of preparation.

B. Increased Sensitivity to Histamine

Routinely we perform this test as follows: Material to be tested is made in three 5-fold concentrations in phosphate-buffered saline, pH 7.2 (plain physiological saline can also be used). For each concentration of material, 10 CFW female mice, 6 to 8 weeks old, are

III. TESTS FOR BIOLOGICAL ACTIVITY

inoculated i.v. with 0.2 ml. Three days later the mice are challenged i.p. with histamine diphosphate in physiological saline containing 0.5 mg histamine base in a 0.2 ml dose. Deaths are recorded 2 h later.

Many unknown factors affect this test and for this reason we carefully standardize and randomize the mice. Male mice are less satisfactory and usually require more pertussigen to become sensitized to histamine.

C. Increased Sensitivity to Serotonin

We perform this test identically to the histamine sensitization test except that 0.5 mg of serotonin (0.5 mg hydroxytryptamine base) in 0.2 ml saline is used for challenge. Serotonin creatinine sulfate is used to make the challenge but the dosage is expressed as free amine. Sensitization to serotonin is affected more strikingly by unknown factors than sensitization to histamine, and the same precautions taken for the histamine sensitization test are followed. A minimum of 10 mice are used per concentration of the test material.

D. Increased Sensitivity to Serotonin-Histamine

This test is identical to B and C except that a mixture of histamine and serotonin is used for challenge. Routinely we use a mixture of 0.05 mg serotonin and 0.2 mg histamine. This is given i.p. in a total volume of 0.2 ml. Some strains of mice have a natural sensitivity to the combinations of these 2 amines and readily die of shock. The CFW is one of these strains.

E. Leukocytosis-Promoting Activity

Five CFW female mice, 5 to 7 weeks old, are inoculated i.v. with the test material contained in a total volume of 0.2 ml phosphate-buffered saline. Three days later blood from the infraorbital plexus is collected into a 40 μl Microcap (Drummond) and immediately expelled into 20 ml of Isoton (Coulter) in an Accuvette (Counter). Six drops of Zap-isoton (Coulter) are added and the leukocytes counted in a Coulter Counter.

Differential counts are made on Wright's or Giemsa-stained blood smears in order to estimate the percentage of lymphocytes, granulocytes, and other cell types.

F. Hypoproteinemia

Pertussigen preparations to be tested are administered i.v. in 0.2 ml of saline to mice. The following day the mice are bled from the infraorbital sinus into plain Caraway (Sherwood) microblood collecting tubes and the sera collected in 100-µl disposable micropipettes. Total serum proteins are determined by the biuret method as described by Annino (3). We have also used a specific gravity method, which dramatically shows differences in total serum protein levels (4), and electrophoresis on cellulose acetate strips (Sepraphore III, Gelman).

G. Hypoglycemia

Pertussigen preparations are administered i.v. in 0.2 ml of saline to mice. Three days later the mice are fasted for 5 h, bled from the infraorbital sinus into heparinized Caraway (Sherwood) microblood collecting tubes, and the plasma collected in 100-µl disposable micropipettes. Plasma glucose is determined by the fluorometric method of Phillips and Elevitch (5).

IV. METHODS TO MEASURE VASCULAR PERMEABILITY

A. Evans Blue Dye Method

The backs of mice are shaved with an electric clipper and then treated with a depilatory (Nair, Carter Products). One to 3 days later the mice receive the materials to be tested intracutaneously via a 30 gauge needle. Saline containing the desired vasoactive substances is injected in a total volume of 0.01 ml. Immediately thereafter the mice receive i.v. 400 µg Evans blue dye in 0.2 ml and 1 h later are sacrificed. The diameters of the blue spots on the inside surface of the skins at the injection sites are measured.

IV. METHODS TO MEASURE VASCULAR PERMEABILITY

B. Radiolabeled Human Serum Albumin (HSA) Method

This is a more quantitative and objective method of measuring vascular permeability. The technique followed is similar to that described by Leibowitz and Kennedy (6). The principle of this technique is that [^{131}I]HSA is distributed throughout the animal's vascular system and gradually leaves the blood. If the vascular bed becomes more permeable, [^{131}I]HSA will leave the circulation at a faster rate. At a certain period after administration of [^{131}I]HSA, a dose of [^{125}I]HSA is given. Five minutes later a blood sample is obtained and then the animal is sacrificed and tissue samples (muscle, liver, brain, etc.) are obtained. By measuring the ^{131}I and ^{125}I activity in the blood and a particular tissue sample, an estimation can be made of the vascular permeability in that tissue. Since most of the [^{125}I]HSA is retained intravascularly during these 5 min, it provides a baseline to measure leakage of [^{131}I]HSA out of the circulation into the extravascular space.

The exact technique is as follows: One microcurie of [^{131}I]HSA (Albumotope inj, Squibb) in saline is given i.v. to each animal. At a later time (1 to 24 h) mice receive i.v. 1 µCi of [^{125}I]HSA (Albumotope diag inj, Squibb) in 0.2 ml saline and 5 min later 20 µl of blood are collected from the infraorbital sinus by means of a calibrated capillary tube (Microcap, Drummond) and the mouse is sacrificed by decapitation. The skin section or tissues to be tested are removed, weighed, and ^{131}I and ^{125}I radioactivity determined in a 2 channel Nuclear Chicago Gamma Scintillation spectrometer. Vascular permeability is calculated and expressed in arbitrary units (extravascular blood equivalents) as described by Leibowitz and Kennedy (6). Briefly, the extravascular blood equivalents (EVBE) are calculated as follows: The specimens are counted for both ^{125}I and ^{131}I activity. The ^{125}I activity is corrected for counts due to ^{131}I "breakthrough" into the ^{125}I channel. The content of ^{131}I and ^{125}I in the tissue is expressed in arbitrary blood equivalent (BE) units for each isotope.

$$BE = \frac{\text{counts per min/g of tissue}}{\text{counts per min/ml of blood}} \times 100$$

Under the experimental conditions ^{125}I is essentially intravascular and ^{131}I both intravascular and extravascular. The difference between $[^{125}I]BE$ and $[^{131}I]BE$ is an indication of how much fluid has leaked out of the vascular bed and gives the extravascular blood equivalent (EVBE = $[^{131}I]BE$ minus $[^{125}I]BE$).

V. DIRECT AGGLUTINATION OF ERYTHROCYTES

To determine the ability of *B. pertussis* extracts to induce agglutination of erythrocytes, we use the following procedure. Either sheep or chicken blood is collected in an equal volume of Alsevier's solution and stored at 4°C for not over 2 weeks. When needed, the erythrocytes are collected by centrifugation and washed 3 times with cold saline. From these washed cells a 0.5% suspension is made.

Serial 2-fold saline dilutions of the extracts are made by means of a Takachi loop in plastic microtiter plates with round bottom wells. To each well 0.025 ml of the erythrocyte suspension is added, mixed, and allowed to stand for 2 h at room temperature and the patterns recorded.

VI. ANTIBODY TITRATION BY THE BIS-DIAZO-BENZIDINE (BDB) TECHNIQUE

The method we use closely follows that described by Arquilla and Finn (7).

A. Preparation of BDB

Benzidine (0.25 g) is dissolved in 45 ml of 0.2 N HCl and cooled in an ice bath. $NaNO_2$ (0.175 g) is dissolved in 5 ml of distilled water and similarly cooled. The cold $NaNO_2$ solution is slowly added with constant stirring (takes 1 min) to the solution of benzidine. The reaction is allowed to continue for 30 min in the ice bath with stirring every 5 min. The reagent can be dispensed in 2-ml aliquots into ampoules, sealed, frozen in an alcohol-dry ice bath, and stored at -15 to -20°C.

VI. ANTIBODY TITRATION BY THE BDB TECHNIQUE

B. Conjugation of Erythrocytes to Antigen

To 3.6 ml of antigen (1 mg HEA per ml) at room temperature, 0.05 ml of 3 times-washed sheep erythrocytes are added and well mixed. Then 0.5 ml of BDB reagent diluted 1/15 in isotonic phosphate buffer at pH 7.4 is added, mixed, and incubated for 10 min at room temperature. The mixture is centrifuged at 1,500 rpm for 5 min and the supernatant removed by decanting. The conjugated cells are resuspended in 5 ml of heated BSA diluent (see buffers below) and recentrifuged as above. The supernatant is decanted and the conjugated erythrocytes are resuspended in 2.5 ml of heated BSA diluent and are ready to use in agglutination test.

C. Preparation of Buffers

Barbiturate buffer containing bovine serum albumin (BDB diluent).

Eighty-five grams of NaCl and 3.75 g of Na-5-diethylbarbiturate are dissolved in 1,400 ml H_2O.

In 500 ml of water, 5.75 g of 5-5-diethylbarbituric acid are dissolved. Then the 2 solutions are mixed, cooled to room temperature, and 5 ml of a stock solution containing 1 M $MgCl_2$ and 0.3 M $CaCl_2$ are added. The total volume is brought up to 2,000 ml with water. This can be stored in the refrigerator. Before use, this stock is diluted 1/5 with water and 1.5 g of BSA (Armour Co.) per liter of the diluted buffer is added. The pH should be close to 7.4. This buffer is then heated at 80°C for 1 h, cooled to 2°C, and is ready for use.

Isotonic phosphate buffer, pH 7.4.

Four parts of 0.104 M Na_2HPO_4 are added to 1 part of 0.104 M NaH_2PO_4 to given an ionic strength of 0.3 and pH 7.35 to 7.4.

D. Hemagglutination Test

Dilutions of the serum to be tested are made in BDB diluent in microtiter plates with U-shaped wells. Standard drop pipets and loops are used to prepare 2-fold serial dilutions. To each well is added a standard drop (0.025 ml) of the conjugated erythrocyte suspension.

The plates are carefully shaken to mix the cells and serum dilutions and then allowed to stand at room temperature for 2 to 3 h and the agglutination patterns recorded.

VII. DETECTION OF ANTIBODY-FORMING CELLS IN SPLEEN AND LYMPH NODES

A. General Remarks

In our hands the most reliable technique to measure antibody-forming cells to a soluble antigen has been that described by Miller and Warner (8) employing chicken gamma globulin as an antigen. The principle of this method is that sheep erythrocytes (SRBD) can be coated with chicken anti-SRBC gamma globulin by merely incubating SRBC with anti-SRBC serum prepared in a chicken. These coated SRBC react with mouse anti-chicken globulin and when guinea pig complement is added, the SRBC lyse. This method is possible because chicken antibody does not fix guinea pig complement and thus the SRBC plus chicken anti-SRBC gamma globulin complex does not lyse cells. The technique we used in our laboratory is as follows:

B. Preparation of Chicken Anti-SRBC Gamma Globulin

Twelve-week-old chickens were immunized i.v. at day 0 with 10^9 SRBC that had been washed 3 times with saline. The chickens were given booster inoculations of SRBC at 10 to 14 day intervals over a 2 month period. Trial bleedings were tested for ability to agglutinate SRBC in microagglutination plates. When the agglutination titer was adequate the chickens were bled out. At a dilution of 1/1,000, a pool of these sera was able to sensitize SRBC.

C. Guinea Pig Complement

A pool of normal guinea pig serum which had been absorbed in the cold twice with an equal volume of packed, 3 times-washed SRBC was used. This complement was stored at -15°C.

VII. SPLEEN AND LYMPH NODE ANTIBODY-FORMING CELLS

D. Sheep Erythrocytes

Sheep blood was collected from the jugular vein directly into an equal volume of Alsevier's solution and stored at 4°C.

E. Balanced Salt Solution (BSS)

(From W. O. Weigle, personal communication.)

Stock solution I:

Dextrose	10 g
Dihydrogen potassium phosphate	0.6 g
Anhydrous disodium phosphate	1.85 g
Phenol red, 0.5% solution	2 ml
Bring volume to 1,000 ml with water	

Stock solution II:

Calcium chloride	1.86 g
Potassium chloride	4.0 g
Sodium chloride	80.0 g
Magnesium chloride	2.0 g
Magnesium sulfate	2.0 g
Bring volume to 1,000 ml with water	

Dilute 100 ml Stock solution I to approximately 800 ml, add 100 ml Stock solution II, and bring final volume to 1,000 ml

F. Amplifying Serum

Guinea pig $7S\gamma_2$ anti-mouse gamma globulin was prepared as follows: Guinea pigs (strain 13) were inoculated with 50 µg purified mouse gamma globulin (Mann, 98% pure mouse gamma globulin) given s.c. in an emulsion of 0.25 ml saline and 0.25 ml Freund's complete adjuvant. Booster doses of antigen were given 14, 22, and 130 days after the primary inoculation in exactly the same manner except that Freund's incomplete adjuvant was used. Sera were obtained and pooled from the guinea pigs at 140 days postprimary inoculation. Guinea pig gamma globulin was separated from the antiserum by a batch procedure which utilized the chloride ion form of diethylaminoethyl (DEAE) Sephadex at pH 6.5 (9). Following this separation, the immunoglobulin fraction was dialyzed against 0.005 M, pH 7.9 phosphate buffer (PB). The proteins were fractionated on a 1.5 x 24 cm column of Whatman DE32 cellulose, which had been previously equilibrated with 0.005 M PB, pH 7.9. The column was eluted with a gradient of PB starting with 0.005 M PB, pH 7.9, and going to 0.3 M PB, pH 7.9. The 1st peak which was eluted from the column was found, when tested by micro-Ouchterlony technique with rabbit antiserum specific for guinea pig immunoglobulins, to contain pure guinea pig IgG_2 immunoglobulin. A pool of these IgG_2 fractions with anti-mouse gamma globulin activity was found to be a good amplifying serum at a 1/100 dilution in balanced salt solution.

G. Collection and Preparation of Lymphocytes

Mouse lymphocytes were obtained by removing spleens and lymph nodes from 3 mice and mincing them through a fine mesh stainless steel screen into a mortar containing about 20 ml BSS. The cell suspension was centrifuged at 1,000 rpm for 10 min and the supernatant fluid removed. Ten milliliters of ammonium chloride (0.83%) was added to the packed cells, mixed, and allowed to stand at room temperature for 5 min. (Note: the ammonium chloride step is omitted for lymph node cells.) The cells were sedimented, the supernate removed, and then the cells were washed with 40 ml of BSS. The cells were sedimented again, washed once more with 10 ml of BSS, and then the cells were suspended in 10 ml of BSS. This was called a 1/10 dilution of cells.

H. Performance of the Test

SRBC in Alsevier's were centrifuged at 2,000 rpm for 5 min, the supernatant fluid removed; then the cells washed twice in saline and then once in BSS. A 5% suspension was made of the washed SRBC in BSS. The 5% suspension of SRBC was incubated with an equal volume of 1/1,000 chicken anti-SRBC gamma globulin for 15 min at 37°C. The sensitized SRBC were washed twice with BSS and then resuspended to a 5% suspension in BSS. A hot solution of 0.5% agarose (L'industrie

Biologique Française S.A.) was dispensed into 12-ml centrifuge tubes (0.5 ml/tube) held in a water bath at 48°C. Microscope slides (25 x 75 mm) were precoated with a 0.2% solution of agarose (made in distilled water) and dried on a slide warmer at 48°C. To the 0.5 ml of agarose in the centrifuge tubes was added 0.05 ml of the SRBC suspension and then 0.1 ml of a mouse lymphoid cell suspension (we usually used a 1/10 and a 1/100 dilution of cells in our tests). The cell mixture was quickly mixed on a vortex mixer and then poured on a warm microscope slide and spread evenly over the slide. The slides were removed from the slide warmer and the agarose allowed to solidify. Slides were then placed with the agarose layer down in special plastic trays and incubated at 37°C for 1 h above the water in a water bath. Plaques were then developed by running a 1/10 dilution of guinea pig complement in BSS under the slides to detect 19 S gamma globulin-producing cells or a BSS solution containing 1/10 complement and 1/100 amplifying serum to detect 7 S gamma globulin-producing cells; both reactions were then incubated for 2 ½ h. Plaques were counted with the aid of a low power hand lens and suitable back lighting. Quantitation was performed by counting the cells in the lymphoid cell suspensions with a hemacytometer and then calculating the number of plaque-forming cells per 10^6 lymphoid cells.

VIII. PASSIVE CUTANEOUS ANAPHYLAXIS (PCA) IN MICE

This test was performed as described by Munoz and Anacker (10).

The backs of mice are shaved with electric clippers. The following day the mice receive an intradermal injection of 0.05 ml of the appropriate antibody dilution and then are challenged 2 h later to measure antibodies of the IgG class or 72 h later to measure the IgE antibodies. The challenge is prepared by making a 0.5% solution of antigen in 0.5% Evans blue dye solution made in physiological saline and 0.2 ml per mouse is given i.v. One-half hour later the mice are killed, skins removed, and the average diameter of the blue spot on the inside surface is measured.

To differentiate between IgG and IgE classes of antibody, one can heat the antibody preparation at 56°C for 3 h. The IgE-PCA activity, which is measured 72 h after sensitization, is destroyed by this treatment while the IgG class, measured 2 h after sensitization, is not destroyed.

IX. ANAPHYLAXIS IN *B. PERTUSSIS*-TREATED MICE

A. Active Anaphylaxis

The procedure we use in mice is similar to that used by Malkiel and Hargis (11). For sensitization, mice receive i.p. the desired amount of antigen mixed with the appropriate adjuvant or *B. pertussis* extract in a total volume of 0.2 ml. Fifteen days later the mice are challenged i.v. with 100 to 500 µg of antigen dissolved in 0.2 ml of physiological saline. Anaphylactic deaths are recorded 3 to 4 h later.

B. Passive Anaphylaxis

The technique is the one described by us previously (12). Mice receive i.p. or i.v. the desired dose of *B. pertussis* cells or extracts in a total volume of 0.2 ml. Three to 4 days after injecting *B. pertussis* cells or extracts, the mice are given i.v. 0.2 ml of the appropriate concentration of antibody. The mice can be challenged 5 to 6 h later with as little as 40 µg of a protein antigen given i.v. in 0.2 ml of physiological saline (routinely we use 200 to 500 µg of HEA).

X. PREPARATION OF ANTISERA

A. Antisera to Purified Proteins

Young adult normal rabbits are used and a preimmunization bleeding of 10 ml is taken. The antigen is dissolved in saline and mixed with an equal volume of complete Freund's adjuvant (Difco) and emulsified by shaking it in a Mickle oscillator for 15 min. Rabbits receive 1 ml of the emulsion containing 200 to 500 μg of antigen per ml in each of the 2 hind footpads (total of 2 ml). Two weeks later the animals receive a s.c. booster dose of 100 μg of the same antigen emulsified in incomplete Freund's adjuvant (0.5 ml antigen solution and 0.5 ml Freund's adjuvant). One to 2 weeks later a trial bleeding is taken and, if the antibody concentration is adequate, the rabbits are bled and the sera are collected. The antisera are kept at -15°C. If the rabbits do not have adequate titers, further booster doses of antigen are given.

B. Antisera to *B. pertussis*

The best technique to immunize rabbits for the production of agglutinins to *B. pertussis* cells is that used by Eldering as described by Kendrick et al. (13).

B. pertussis is grown in B-G agar for 36 to 7s h at 37°C. About 3 ml of saline is added and the growth is carefully removed by means of a bent glass capillary. The cell suspension is passed through a double layer of gauze and standardized to contain 10×10^9 cells per ml. The cells are killed by 1/10,000 merthiolate and incubated for at least 48 h at 4°C.

After a preimmunization bleeding, rabbits are inoculated i.v. with 4 doses of this vaccine given at 3- to 4-day intervals. These doses are 0.2, 0.4, 0.8, and 0.8 ml per kg body weight. One week after the last injection a trial bleeding is made. If the agglutinin titer against the homologous strain is 4,000 or higher, the rabbits are bled and the sera pooled. If the sera are unsatisfactory, further immunization may increase the titers.

XI. AGGLUTINATION TESTS

For this purpose we collect the cells as we do for immunization, except that they are suspended in 0.3% formalin. These cells form a more stable smooth suspension and are more satisfactory for performing agglutination tests.

Serum dilutions to be tested are made in saline employing either tubes or microagglutination plates. An equal volume of the suspension of cells containing about 10×10^9 cells per ml are added to each tube or well and mixed. If tubes are used, we incubate them at 56°C for 1 h and then overnight in the cold and read the results the following day. Microagglutination plates are allowed to stand at room temperature for 3 h before the results are read.

For typing cultures, the slide agglutination test is used. One drop of a fresh (24- to 36-h-old culture) cell suspension (40×10^9 cells per ml) is mixed with 1 drop of the appropriate dilution of the specific typing serum. The cells are mixed well and the slide or plate rocked several times and incubated 5 min in a partially closed box to prevent evaporation. The slide or plate is rocked again and the agglutination read.

XII. AGGLUTININ PRODUCTION IN MICE

To test for the presence of agglutinogens in fractions of B. pertussis cells, we inoculate mice, in groups of 3 to 5, with 2 doses of the fraction, given i.p., 2 weeks apart, and bleed them 2 weeks after the 2nd injection. The bloods of the 3 to 5 mice are pooled, the serum collected, and the agglutinin content titrated by microagglutination in plastic plates. The usual doses of whole cells range from 0.2×10^9 to 5×10^9 per mouse, while with fractions doses range from 5 to 125 µg.

XIII. AGGLUTINOGEN ABSORPTION TEST

This test is used to determine the amounts of agglutinogen present in various fractions. It is performed as follows:

Saline solutions of antigen are mixed with an equal volume of antiserum diluted to a concentration twice its plate agglutinin titer. The mixture is incubated 15 min at room temperature, mixed again, and allowed to incubate an additional 15 min. The mixture is centrifuged and the supernatant serum tested for agglutinins by the plate or tube agglutination test. Nonspecific absorption is controlled by using extracts free of agglutinogens.

XIV. IMMUNIZATION OF MICE TO PRODUCE ASCITIC FLUID CONTAINING IgE ANTIBODIES

In our work with the production of IgE type of antibodies to HEA in mouse peritoneal fluid, the following procedure outlined in Table 1 has proven successful in both C57BL/6J and CFW mice. High titers of IgG antibody are also obtained in this fluid.

TABLE 1

Protocol for the Production of Ascites Containing High Titer of IgE-Like Antibodies to HEA

Day of treatment	HEA (µg/mouse)	BPE (µg/mouse)	Freund's complete adjuvant (ml/mouse)
0	125	50	--
7	--	--	0.5
14	--	--	0.5
21	--	--	0.5
28	5	--	0.5
35	5	--	0.5
47	5	--	0.5
56	Collect ascites	--	0.5

The HEA and BPE are given mixed in a volume of 0.2 ml of phosphate buffered saline and given i.p.

The ascites starts to develop shortly after the 2nd injection of complete Freund's.

The mice are tapped as they develop marked ascites. The IgE titers to HEA of this fluid are as high as 1/1,000 in many cases.

XV. ENHANCEMENT OF EXPERIMENTAL ALLERGIC ENCEPHALOMYELITIS

The spinal cord of adult Hartley strain guinea pigs is removed and emulsified in physiological saline (100 mg, wet weight, cord per ml). This emulsion is heated at 60°C for 45 min and emulsified again by passing repeatedly through a 19 gauge needle (14). This material is stored at -15°C.

Female Lewis strain rats, weighing from 100 to 125 g, are given i.p. 200 mg (wet weight) of guinea pig spinal cord antigen mixed with the appropriate amount of *B. pertussis* fraction. Three dose levels of each fraction are tested, and 3 to 5 rats are used per dose of fraction. A similar group of 3 to 5 rats receive i.p. 200 mg of cord emulsion alone. The rats are observed daily for signs of paralysis. To confirm EAE histologically, spinal cord is removed, fixed in 10% buffered formalin and sections cut, stained with hematoxylin-eosin, and examined under the microscope. In hyperactute EAE a marked perivascular accumulation of monocytic cells mixed with polymorphonuclear lymphocytes occurs. Accumulation of fibrin is also noted.

XVI. SCHULTZ-DALE REACTION

This reaction is carried out as described by Munoz and Maung (15). The uteri of mice that have just been killed are carefully dissected free and placed in Tyrode's solution of the following composition:

Sodium chloride	8 g
Potassium chloride	0.2 g
Calcium chloride	0.05 g
Magnesium chloride	0.1 g
Sodium dihydrogen phosphate	0.05 g
Sodium bicarbonate	1 g
Glucose	1 g
Distilled water	1,000 ml

pH of this solution is 7.6

XVIII. IMMUNOELECTROPHORESIS   211

The solution is kept overnight at room temperature before use. One of the uterine horns is connected to an isotonic myograph and the muscular contractions are recorded with an E & M Physiograph. When the muscle has completely relaxed, it is challenged with the antigen (HEA) in a concentration of 0.05 mg per ml of bath fluid (the best concentration has to be determined for each antigen). The viability of muscles not responding to antigen is tested by adding 0.05 µg of serotonin or 10 µg of acetycholine per ml of bath fluid.

The uteri can be stored 24 h at 2 to 5°C without affecting their reactivity.

XVII. GEL DIFFUSION TEST

Routinely this test is performed by covering clean microscope slides with a layer of melted 1% agarose (3 ml) made in saline, allowing it to solidify, and then cutting wells with a sharp hollow punch inserted through holes in a Lucite template. The agarose plugs are removed by vacuum applied to a capillary pipet. The antigen and antibody solutions are placed in the desired wells, incubated at room temperature in a moist chamber for 24 h, and read against a dark background with oblique backlighting. The slides are photographed for permanent records.

XVIII. IMMUNOELECTROPHORESIS (IEP)

Routinely this test is performed on standard clear microscope slides that have been precoated with 1% agarose and dried in an oven at 120°C for a few minutes. Three milliliters of melted 1% agarose in Veronal buffer (μ = 0.039, pH 8.2) are layered on each slide and allowed to solidify. Wells and troughs are cut with a commercial cutter (Gelman Instruments Co., Ann Arbor, Mich). The agarose plugs in the wells are removed by applying vacuum with a capillary pipet and the sample to be electrophoresed is applied. The slices are placed in an electrophoresis apparatus and paper wicks placed in each buffer vessel which contain Veronal buffer. The apparatus is designed so that when in operation little or no evaporation occurs. Crushed ice is also placed around the buffer vessels. Thirty to 35 volts across the length of the slide is applied for 1 to 2 h, depending on the mobility of the antigen. At the end of this period the electricity is turned off, the slides are removed, and the agarose in the troughs removed by means of vacuum through a capillary pipette. The troughs are filled with antiserum and the slides incubated in a moist chamber for 24 h at room temperature. At the end of this period the slides are examined by indirect illumination in a special viewer and photographed for permanent records.

XIX. DISC ELECTROPHORESIS IN ACRYLAMIDE GEL

The apparatus used for this was purchased from Buchler Instruments, Inc., Fort Lee, N. J. This apparatus consists of an upper buffer chamber and a lower buffer chamber. The upper chamber has holes with rubber O-rings to make a water-tight fitting around the electrophoresis tubes. The cathode is placed in the upper chamber and anode in the lower. The ends of the tubes containing the gel and sample make contact with upper and lower buffers. The procedure followed was that described by the manufacturers. The electrophoresis tubes are capped on one end and placed in a holding rack. The tubes are filled to a depth of 4 cm with the lower gel (Table 2) and then a layer of 4 mm of water placed over the gel to ensure a smooth flat surface on the gel. The gel is photopolymerized for 30 min by placing a fluorescent light 3 inches behind the tube rack. The water is carefully aspirated, the surface of the gel is rinsed with 1 drop of the upper gel and removed by aspiration. Add 0.15 ml of the upper gel (Table 2) to each tube and then carefully overlay with 4 mm of water. The upper gel is photopolymerized for 15 min and the water removed by aspiration. The rubber caps are removed from the ends of the tubes and the tubes are placed in position in the upper buffer chamber. Fill the lower chamber with the lower buffer (Table 2) to a depth of 10 cm and place a drop of buffer on the lower end

XIX. DISC ELECTROPHORESIS IN ACRYLAMIDE GEL

TABLE 2

Materials Used for Disc Electrophoresis

Anionic gel system, running pH 9.3			
	Solution		
Lower gel	a	Acrylamide	30 g
		Bisacrylamide	0.8 g
		Water, q.s.	100 ml
	b	Tris (Trizma base)	18.15 g
		1 N HCl	24 ml
		Temed (N,N,N',N' tetra-methylethylenediamine)	0.24 ml
		Water, q.s.	100 ml
	c	Ammonium persulfate	0.14 g
		Water, q.s.	100 ml
		Mix 1 part a, 1 part b, and 2 parts c	
Upper gel	d	Acrylamide	10 g
		Bisacrylamide	0.8 g
		Water, q.s.	100 ml
	e	Tris	2.23 g
		1 M H_3PO_4	12.8 ml
		Temed	0.1 ml
		Water, q.s.	100 ml
	f	Riboflavin	2 mg
		Water, q.s.	100 ml
	g	Ammonium persulfate	80 mg
		Water, q.s.	100 ml
		Mix equal parts d, e, f, and g	
Upper buffer		Tris	5.16 g
		Glycine	3.48 g
		0.001% bromphenol blue in H_2O	2 ml
		Water, q.s.	1000 ml
Lower buffer		Tris	14.5 g
		1 N HCl	60 ml
		Water, q.s.	1000 ml

Electrophoresis tubes: glass 75 mm L. x 5 mm ID x 8 mm OD.

Destaining tubes: glass 75 mm L. x 6 mm ID x 8 mm OD one end slightly constricted.

Clean tubes with acid cleaning solution, rinse with tap H_2O, then distilled H_2O; finally with a dilute Kodak Photo-Flow 200 solution. Dry; repeat before each use.

of each tube before lowering them into the buffer (this displaces any air bubbles on the lower end of the gel). Place the upper chamber on top of the lower chamber so that the anode end of the tubes makes contact with the lower buffer, and fill the upper chamber with upper buffer. The lower chamber is kept cold by circulating cold water through a water jacket surrounding the chamber. Fifty microliters of a sample, made heavy by adding 5 to 10% sucrose, is applied on top of each tube. (The upper gel layer should always be at least twice the size of the sample.) Now place the cover with the electrodes in place and make sure the apparatus is level and connect the power supply. The positive lead goes to the lower electrode and the negative to the upper. A current of 1.25 ma per gel column is applied until the sample has entered the stacking (upper) gel (a blue band appears in the gel; this is the bromphenol blue dye which follows the protein bands). At this point the current is increased to 2.5 ma per column for the remainder of the run. The leading buffer front marked by the dye is followed visually and the run is terminated when the dye reaches a point about 3 to 5 mm from the bottom of the gel (in the lower gel, the dye overtakes and precedes the protein bands). The power is shut off and the gel columns removed from the tubes by gently ringing the gel column with a long 22 gauge needle through which water is being forced. The gel is cut off at the buffer front (dye band) for a reference point and placed in a 10 x 75 mm tube. The gel is now covered with 0.5% amido Schwarz in 7% acetic acid solution to fix and stain the protein bands. The gel is stained for 60 min. Now the gels are placed in destaining tubes with the tapered end of the tube down. The tubes are placed in the upper chamber, the lower chamber is filled with 7% acetic acid, and the upper chamber is placed in position as previously described. The upper chamber is filled with 7% acetic acid. Make sure all air bubbles under or around the gels are removed. Connect the power supply as before and apply a current of 5 ma per column until all excess dye is removed from the columns. The gel may be stored in closed vials in 7% acetic acid. The gels are best viewed or photographed while back lighted with a fluorescent light.

XX. HYDROXYLAPATITE COLUMN CHROMATOGRAPHY

Hydroxylapatite (BioGel HTP) was purchased from Bio-Rad Laboratories, Richmond, Calif. Phosphate buffers were all made with NaH_2PO_4 and Na_2HPO_4 at pH 6.8. Only the molarities were varied; 0.01 M, 0.1 M, 0.2 M, and 0.65 M were used.

To prepare the column, 50 g of hydroxylapatite is washed twice with 0.01 M buffer. The slurry is poured into a K25/45 LKB Sephadex laboratory column equipped with upward flow adaptors at each end. The hydroxylapatite is allowed to settle by gravity and then buffer is run through to pack the column. The upper flow adaptor is placed so that the bottom part just touches the hydroxylapatite and then 0.01 M buffer passed overnight. The column is connected to a Buchler Fractomat automatic fraction collector with the effluent tube passing through an LKB Uvicord II UV absorptiometer for continuous monitoring and recording of the UV absorption; recorded as percent transmittance. Crude B. pertussis extract (150 ml) in 0.01 M buffer is pumped onto the column and then un stepwise manner the column is eluted with 0.1 M, 0.2 M, and 0.65 M buffers to elute various fractions. This method elutes most of the HSF in the 0.2 M fraction. The 0.1 M buffer is added immediately after the sample; the 0.2 M buffer is started when the percent transmittance reaches 90; the 0.65 M buffer is started also when the percent transmittance is again about 90. The flow rate of the buffers through the column is controlled by the pump at a rate of 1.2 ml per min. A fraction is cut every 15 min.

XXI. STARCH BLOCK ELECTROPHORESIS

Purified potato starch (Baker) is washed several times with distilled water and then with phosphate buffer ($\mu = 0.025$, pH 6.2). The slurry is poured into a mold 52 cm long, 6 cm wide, and 1 cm deep lined with vinylidene chloride-vinyl chloride copolymer film (Saran wrap). The starch slurry is blotted with filter paper until a firm consistency is obtained. The sample is applied in a starch slurry to a 1 x 2 x 6 cm trough cut out of the starch block near the center of the block, and making sure that good contact is made between the starch of the block and the sample. After application of the sample, the block is wrapped with Saran wrap except the ends where sponge wicks are placed to make contact with buffer vessels. Electrophoresis is carried out for 46 h at 10 ma and 310 volts. At the end of the run the block is cut into 2-cm fractions which are extracted with 20 ml of buffer pH 7.2 containing 0.5 M NaCl.

XXII. ZONAL DENSITY GRADIENT ELECTROPHORESIS

The zonal density gradient electrophoresis is performed as described by Svensson (16) in an LKB Porath electrophoresis apparatus. A linear sucrose gradient (10 to 40%) is made in 0.03 M KH_2PO_4 buffer at pH 7.8 (12.3 g KH_2PO_4 + 82.4 ml 1 N NaOH in total volume of 6 liters). On top of the gradient a 4 ml sample of AP is layered on top of the gradient and an electric current of 16 ma and 600 volts applied for 14 to 16 h. Two hundred fifty drop fractions (about 10 ml) are collected by means of a Buchler Fractomat collector. A constant recording of OD at 280 nm is taken by a LKB Uvicord recorder. Each fraction is tested for presence of antigens by an hyperimmune serum made against 0.2 M fraction and also by adding trichloroacetic acid to detect precipitable material. In addition, the fractions are tested for their ability to sensitize mice to histamine.

REFERENCES

1. S. W. Cohen and M. W. Wheeler, Amer. J. Public Health, 36, 371 (1946).
2. M. Pittman and J. E. Lieberman, Amer. J. Public Health, 38, 15 (1948).
3. J. S. Annino, Clinical Chemistry; Principles and Procedures, 3rd ed., Little, Brown, and Co., Boston, 1964, p. 184.
4. O. H. Lowry and T. H. Hunter, J. Biol. Chem., 159, 465 (1945).
5. R. E. Phillips and F. R. Elevitch, Amer. J. Clin. Pathol., 49, 622 (1968).

REFERENCES

6. S. Leibowitz and L. Kennedy, *J. Immunol.*, *22*, 859 (1972).
7. E. R. Arquilla and J. Finn, *J. Exptl. Med.*, *122*, 771 (1965).
8. J. F. A. P. Miller and N. L. Warner, *Int. Arch. Allergy*, *40*, 59 (1971).
9. J. S. Baumstark, R. J. Laffin, and W. A. Bardawil, *Arch. Biochem. Biophys.*, *108*, 514 (1964).
10. J. Munoz and R. L. Anacker, *J. Immunol.*, *83*, 640 (1959).
11. S. Malkiel and B. J. Hargis, *J. Allergy*, *23*, 352 (1952).
12. J. Munoz, L. F. Schuchardt, and W. F. Verwey, *J. Immunol.*, *80*, 77 (1958).
13. P. L. Kendrick, H. E. Alexander, W. L. Bradford, G. Eldering, and M. Pittman, *Diagnostic Procedures and Reagents* (A. Harris and M. Coleman, eds.), 4th ed., American Public Health Association, Inc., New York, 1963, p. 398.
14. S. Levine, E. J. Wenk, H. B. Devlin, R. E. Pieroni, and L. Levine, *J. Immunol.*, *97*, 363 (1966).
15. J. Munoz and M. Maung, *Proc. Soc. Exptl. Biol. Med.*, *106*, 70 (1961).
16. H. Svensson, in *Protein Chemistry Including Polypeptides* (P. Alexander and R. J. Block, eds.), Vol. 1, Pergamon Press Ltd., London, 1960, p. 193.

Author Index

Numbers in parentheses are reference numbers and indicate that an author's work is referred to although his name is not cited in the text. Italicized numbers give the page on which the complete reference is cited.

A

Abernathy, R. S., 49(104), 69, 73(29), *104*
Abrahams, I., 169, *182*
Adler, A., 43(77), 50(77), 53 (77), *68*, 156(19), *157*, 187(21), *190*
Ahearn, M. J., 170(26), *182*
Alexander, H. E., 207(13), *217*
Allison, A. C., 129(20), *140*
Anacker, R. L., 26(30), 27(30, 32), 28(32), 29(30), 49 (109), *66, 69*, 73(12), 93 (12), *103*, 115(19), 116 (19, 24, 25), 117(25), 118(24), *121, 122*, 205, *217*
Andersen, E. K., 32, *67*, 166 (16), 180,*181, 182*
Anderson, H. R., 128(15), 129 (15), 139(15), *140*
Andreesco-Tigoiu, V., 151(3), *157*
Annino, J. S., 198, *216*
Arai, H., 43(76), 45(89), 49 (76), 50(89), 51(76, 111), 55(76, 89, 111, 60), 64, *68, 69*, 153(14), 157(14), *157*, 188(23), *190*
Arch, R. N., 73(24), *104*, 165, *181*
Arquilla, E. R., 200, *217*
Asakawa, S., 51(118), 62(118), *70*, 135(24, 26), 136(26), 139(26), *140* 153(15), 157 (15), *157*

Asherson, G. L., 139(28), *141*
Askonas, B. A., 129(20), *140*
Attig, C., 176(39), 177(39), *182*

B

Baker, J., 32(41), 38(57, 59), 39(66), *67, 68*
Banerjea, A., 21(13), 22(13), *65*
Bardawil, W. A., 203(9), *217*
Barron, B. A., 154(18), 156(18), *157*, 187(19), *190*
Bartoschek, M., 127(13), 139 (13), *140*
Baumstark, J. S., 203(9), *217*
Beith, E. M., 47(99), *69*
Bell, J. F., 166(21), 167(21), 168(21), 169(21), 170 (21), *182*
Beneke, G., 128(17), *140*
Benner, M., 139(30), *141*, 186 (12), *190*
Berenbaum, M. C., 187(17), *190*
Bergman, R. K., 9(34), 10(34), *12*, 31(36, 37), 39(60), 53(36), *66, 67*, 73(11), 74(48, 55), 76(48, 55, 56), 77(11, 48), 80(11), 81(67), 82(11), 84(67), 85(11), 86(11), 87(67), 88(90), 89(90), 90(96, 97), 91(97, 98), 92(97), 94(98, 114, 116), 98 (117), 99, 100(122), 101 (124), *103, 105, 106,*

[Bergman, R. K.]
  107, 116(26), 119(29, 30, 31), 122, 124(4, 8), 125(4), 126(8), 130(4, 8), 132(4), 135(4), 140, 148(11), 150, 151(5), 153(7, 10), 157(5), 157, 160(6), 161(6), 162(6), 163(10), 164(10), 181, 184(2, 5, 6, 7), 185(8), 189
Berkelhammer, J., 148(12, 13), 150
Bernard, C. A., 145, 150
Bernstein, I., 171(31), 182
Binaghi, R., 130(21), 131(21), 140
Biozzi, G., 171(30), 182
Bister, F., 26(29), 66, 126(11), 140
Blaskett, A. C., 39(65), 67
Blum, L., 45(85), 68
Bogaerts, W. J. C., 169, 182
Boivin, A., 26, 66
Bondi, A., Jr., 26(25), 38(52), 66, 67
Bordet, J., 1(3, 4, 5), 10, 11, 20, 25(9), 32, 43(80), 65, 66, 68
Borecký, L., 173(36), 182
Boussac-Aron, Y., 130(21), 131(21), 140
Boyer, F., 73(31), 93(112), 104, 107
Bradford, W. L., 1(1), 10, 145(8), 150, 151(2), 157, 207(13), 217
Brandon, E. M., 110(5), 121
Bray, K. K., 54(119), 60, 70, 153(11), 154(11), 156(11), 157(11), 157
Broderick, E. J., 31(34), 45(88), 54(88), 66, 68, 76(51), 105
Bronne-Shanbury, C. J., 38(58), 67
Brown, R., 26(30), 27(30), 29(30), 66
Brüss, E., 128(14), 140(14), 140, 186(10), 190
Burnet, F. M., 9(28), 11
Byers, R. K., 8, 11, 22(19), 66

C

Cambasse, H., 21(14), 65
Cameron, C., 18(6), 19(6), 27(6), 30(6), 65
Cameron, J. J., 34, 67
Carnegie, P. R., 145, 150
Carr, R. H., 139(30), 141, 186(12), 190
Chalvardjian, N., 39(64), 67
Char, D. F. B., 185(9), 190
Chedid, L., 73(30, 31), 93(103, 112), 104, 107
Chutkow, J., 176(39), 177(39), 182
Ciplea, A. G., 151(3), 157
Clausen, C. R., 124(4, 8), 125(4), 126(8), 130(4, 8), 132(4), 135(4), 140, 153(7), 157
Cohen, H., 45(94), 69
Cohen, S. W., 193, 216
Cole, R. L., 51(113), 69, 111(12), 113(12), 121, 175(38), 182
Corley, E. J., 169(24), 182
Craft, C. E., 90(95), 106
Craft, K., 170(27), 182
Cronholm, L. S., 89, 98, 106, 107
Csaba, B., 88(83), 98(83), 106, 160(5), 181
Csongor, J., 88(83), 98(83), 106, 160(5), 181

D

Dagnelie, J., 22(18), 66
Davis, A., 39(66), 68
Dechene, E., 24(23), 66
Denman, P. M., 139(28), 141
Devlin, H. B., 50(112), 69, 145(5), 146(5), 147(5), 150, 210(14), 217
Dietrich, F. M., 113(16), 121
Dixon, M. K., 9(29), 12, 38(54), 48(54), 67
Dolby, J. M., 38(58), 67, 186(13), 190
Douglas, W. H., 91(101), 107

AUTHOR INDEX

Dresser, D. W., 128, 129(15), 139(15, 31, 34), 140(36), 140, 141, 186(11), 190
Dubos, R. J., 73(25), 104, 166, 182
Durville-van der Oord, B. J., 169, 182

E

Ehrich, W. E., 26, 66
Eldering, G., 2, 9(29), 11, 12, 26, 32, 33(42, 43), 34(42), 38(54, 57), 39(66), 44, 48(54), 66, 67, 68, 72, 73(26), 103, 104, 110, 121, 207(13), 217
Elekes, E., 129(19), 140
Elevitch, F. R., 198, 216
Emmerling, P., 127(13), 128(14, 16), 139(13), 140(14), 140, 186(10), 190
Enta, T., 139(30), 141, 186(12), 190
Evans, D. G., 21(15), 22(15), 66
Evans, O. G., 166(17), 182
Eveland, W. C., 33(42), 34(42), 67

F

Farthing, J. R., 30(33), 66, 126, 127(9), 140
Faur, Y., 151(3), 157
Feinberg, S. M., 81(66), 105
Felton, H. M., 38(52), 43(81), 44(84), 67, 68
Festenstein, H., 139(28), 141
Finger, H., 127, 128(14, 16, 17, 18), 129(19), 139, 139(14), 140, 186(10), 190
Fink, M. A., 73(21), 91, 104, 120(32), 122
Finn, J., 200, 217
Fishel, C. W., 76(49), 88(84, 87, 88), 89(84), 94, 96(49), 97(115), 98, 105, 106, 107, 150(19), 150, 160(8, 9), 181, 184, 159

Fisher, S., 42(71, 74), 68
Fleming, D. S., 47(98, 99), 69, 123, 124(2), 139
Floersheim, G. L., 170(25), 171, 173(25), 182
Flosdorf, E. W., 26(25), 35, 38, 66, 67
Földmer, I., 128(18), 140
Fonteyne, P., 22(18), 66
Food and Drug Administration, 45(92), 68
Fox, C. L., 110(4), 121
Fresenius, H., 128(17), 140
Freund, J., 148, 150
Fröhlich, J., 7, 11, 151, 157
Fukumura, K., 154(16), 155(16), 156(16), 157, 187(20), 190
Fukushi, K., 27(32), 28(32), 66
Fulton, J. D., 90(95), 106
Fulton, L. C., 39(55), 67

G

Gadsden, R. H., 49(102), 69, 73(38), 104
Gallavan, M., 22(17), 66
Ganley, O. H., 88(79, 80), 89(93), 94(93), 106, 160, 181
Ganong, W. F., 72(1), 103
Gardner, A. D., 2, 11, 18(4), 65
Gauthier, G. F., 76(57), 105
Geller, B. D., 82, 105
Gengou, O., 1(3, 4), 10, 20, 25(9), 43(80), 65, 68
Germuth, F. G., 49(108), 69, 73(14), 103, 115, 122
Gershon, M. D., 73(22), 104
Gershon, R. K., 139(35), 141
Glick, M., 170(27), 182
Golub, O. J., 115(17), 121
Goodline, M. A., 9(33), 12, 48(100), 49(100), 69, 72(4, 5, 6), 73(4, 5, 6, 16), 103, 110(9), 121
Goodpasture, E. W., 22(17), 66
Gözsy, B., 88(89), 106, 160(3), 181

Grasso, A. Y., 88(85), *106*
Greenberg, L., 47(98, 99), *69, 123, 124(2), 139*
Gruenewald, R., 177, *182*
Guerault, A., 73(45), 77, *104, 105*
Gulasekharam, J., 39(65), *67*
Gulbenkian, A., 9(35), *12*, 88 (82, 85), 98, *106*, 160(4), *181*

Ishida, S., 51(118), 62(118), *70*, 135(24), *140*, 153(15), 157(15), *157*
Ishizaka, I., 131(22), *140*
Ishizaka, T., 131(22), *140*
Iwasa, S., 51(118), 62, *70*, 135 (24, 25), 137(27), 139(27), *140, 141*, 153(15), 154(16), 155, 156(16), 157(15), *157*, 187(20), *190*

H

Halpern, B. N., 91(99), 93(99), *107*, 171(30), *182*
Hamada, K., 87(75), *106*
Hambrecht, L., 7, *11*
Hamre, D. M., 115(17), *121*
Hannik, C. A., 8(22), *11*
Hargis, B. J., 31(35), *66*, 73 (10, 41, 43), 76(54), 77 (60), 81(66), *103, 104, 105*, 110(2, 13), *121*, 148 (12, 13), *150*, 173(34), *182*, 206, *217*
Harris, W. E., 90(95), *106*
Hasenclever, H. F., 169(24), *182*
Haskins, W. T., 26(30), 27(30, 32), 28(32), 29(30), *66*
Henseling, M., 128(18), *140*
Hestekin, B. M., 45(86), 46(97), 50(97), 54(86), *68, 69*
Hiramatsu, T., 51(117), *70*
Hirano, M., 170(26), 171, *182*
Hof, H., 129(19), *140*
Holt, L. B., 30(33), 35, *66, 67*, 126, *140*
Holwerda, J., 38(57, 59), 39(66), *67, 68*
Hornbeck, C., 32(41), *67*
Hornor, A., 7(20), *11*
Howe, G. D., 171(28), *182*
Hunder, G., 73(40), *104*
Hunter, T. H., 198(4), *216*

J

Jakus, C. M., 51(115), *69*
Jenkins, H. J., 76(57), *105*
Joo, I., 46(96), *69*, 82(68), *105*
Juhász, V. P., 82(68), *105*

K

Kabat, A., 26, *66*
Kallos, P., 49(106), *69*, 73(20), *104*
Kallos-Deffner, L., 49(106), *69*, 73(20), *104*
Kátó, L., 88(89), *106*, 160(3), *181*
Katsh, S., 88(81), *106*
Keller, K. F., 88(88), *106*, 160(9), *181*
Kelley, V. C., 185(9), *190*
Kendrick, P. L., 2, 9(29), *11, 12*, 33(42), 34(42), 38 (54), 44, 48, *67, 68*, 207, *217*
Kennedy, L., 149, *150*, 199, *217*
Keogh, E. V., 39, *68*
Kimball, A. C., 35, *67*
Kind, L. S., 49(102, 105), *69*, 73(19, 32, 38), 74(59), 75, 76(19, 53), 77(59), 87(71), 91(19), 93(113), *103, 104, 105, 107*, 110 (14), *121*, 166(14), *181*
Kishigami, M., 87(72, 73), *105*
Kitagawa, M., 20(11), 36(51), 38(51), 51(11), *65, 67*
Klein, T. W., 100(120), *107*
Kohn, R., 74(46), 75(46), 81 (46), *104* 153(8), *157*

I

Iida, T., 20, 21(12), 23, *65*, 166(18), *182*, 186, 187(18), *190*

AUTHOR INDEX

Kojima, Y., 173(35), 174(35), 182
Komarek, A., 113(16), 121
Konosu, M., 187 (18), 190
Kruse, H., 90, 106
Kuo, C. Y., 173(33), 182
Kurokawa, M., 51(118), 62(118), 70, 135(24), 140, 153(15), 154(16), 155(16), 156(16), 157(15), 157, 187(20), 190
Kusano, N., 187(18), 190
Kuwajima, Y., 87(72, 73, 74, 75, 76, 77), 105, 106

L

Lackovič, V., 173(36), 182
Laffin, R. J., 203(9), 217
Landy, M., 27(32), 28(32), 66, 166, 181
Lapin, J. H., 7(16), 8(16), 11, 151(4), 157
La Pointe, D., 24(23), 66
Larson, C. L., 14(2), 15(2), 20(2), 22(2), 50(2), 65
Lawson, G. M., 14(1), 65
Laxson, C., 116(23), 122
Lee, J. M. 143, 150
Lehrer, S. B., 45, 51(90, 114), 55(120), 60, 68, 69, 70, 157(22), 158
Leibowitz, S., 101, 107, 149, 150, 199, 217
Lepow, I. H., 45(85), 68
Leslie, P. H., 2, 11, 18(4), 65
Levine, L., 31(34), 45(88), 50 (112), 54(88), 66, 68, 69, 73(35, 36, 42), 76(42, 51), 104, 105, 145(5), 146(5), 147, 150, 210(14), 217
Levine, S., 50(112), 69, 144(2), 145(4, 5, 6, 7, 9), 146 (5, 9), 147(10), 148(15), 150, 175, 176(37), 177, 182, 210(14), 217
Lieberman, J. E., 9(30), 12, 38 (55), 67, 195, 216
Likhite, V. V., 171, 172(32), 182

Lockey, S. D., 139(30), 141, 186(12), 190
Loew, E. R., 76(57), 105
Lowry, O. H., 198(4), 216
Lüderitz, O., 26(29), 66, 126, 140

M

MacDonald, A. D., 74(46), 75 (46), 81(46), 104, 153(8), 157
McGarry, B., 88(81), 106
McGuinness, A. C., 38(52), 67
MacLennan, A. P., 26, 27, 29 (28), 66
McMaster, P. D., 90, 106
Maitland, H. B., 21(15), 22(15), 66, 74(46), 75, 81(46), 104, 153(8), 157
Makino, S., 139(30), 141, 186 (12), 190
Malkiel, S., 31(35), 66, 73(10, 41, 43), 76(54), 77(60), 81(66), 93(109), 103, 104, 105, 107, 110(2, 13), 121, 148, 150, 173(34), 182, 206, 217
Mallory, F., 7(20), 11
Malmgren, B., 26(30), 27(30, 32), 28(32), 29(30), 66
Marmorston, J., 93(106), 107
Masry, F. L. G., 40(70), 41, 42 (70), 43(70), 68
Matsui, T., 87(72, 73, 76), 105, 106
Matsumura, Y., 98, 107
Maung, M., 118(28), 119(28), 122, 210, 217
Medical Research Council, 9(31, 32), 12, 38(56), 44(56), 67
Mesrobeanu, I., 26(27), 66
Mesrobeanu, L., 26(27), 66
Michael, J. G., 73(33), 104
Miles, A. A., 9(27), 11, 110(1), 121
Miller, J. F. A. P., 202, 217
Milner, K. C., 26(30), 27(30), 28(32), 29(30), 66

Misner, J., 9(29), *12,* 38(54), 48(54), *67*
Mohr, W., 129(19), *140*
Moll, F. C., 22(19), *66*
Moos, W. S., 73(37), *104*
Moreno-López, M., 2, *11*
Morgan, P., 110(6), *121*
Morse, J. H., 43, 49, 51(79), 54, 60, 64, *68,* 153(13), 157(13), *157*
Morse, S. I., 43(77), 45, 49, 50(77), 51(79), 53(77), 54, 60, 64, *68, 70,* 100 (121), *107,* 151, 152(6), 153(6, 11, 12, 13), 154 (6, 11, 17, 18), 155(6), 156(6, 11, 18, 19, 20), 157(11, 12, 13), *157,* 184(3), 187(19, 21), *189, 190*
Mota, I., 124(5, 6, 7), 130, 132(23), *140*
Mouton, D., 171(30), *182*
Mudd, S., 26(25), 35, *66, 67*
Muggleton, P. W., 8(26), *11*
Munoz, J. J., 9(34), 10(34), *12,* 14(2), 15(2), 18(5, 6, 7), 19(5, 6), 20(2, 10), 21 (10, 13), 22(2, 13), 27 (6), 30(6), 31(36, 37), 32(5, 39), 33(44), 34(5), 35(39), 36(39), 38(39), 39(60), 42(7), 45(7, 86), 46(97), 47(7), 48(5), 49 (103, 107, 109), 50(2, 97), 51(5, 113), 53(5, 36), 54(86), 56(7), *65, 66, 67, 68, 69,* 73(11, 12, 13, 15, 17, 39), 74(48, 52, 55), 75(63, 65), 76 (17, 48, 52, 55, 56), 77 (11, 48), 79(65), 80(11), 81(63, 67), 82(11), 84 (67), 85(11), 86(11), 87 (67), 88(90), 89(90), 90 (96, 97), 91(97, 98), 92 (97), 93(12, 39), 102, 104, 105, 108), 94(98, 114, 116), 95(98), 96(98), 97 (116), 98(117), 99, 100 (104, 122), 101(124), *103,*

[Munoz, J. J.]
104, 105, 106, 107, 111 (12), 113(12), 114(15), 115(19, 21, 22), 116(19, 22, 23, 24, 25, 26), 117 (25), 118(21, 22, 24, 27, 28),, 119(28, 29, 30, 31), *121, 122,* 123(3), 124(4, 8), 125(4), 126(8), 130(4, 8), 132(4), 135(4), 139(29), *140, 141,* 148(11, 17) *150,* 151(5), 153(7, 9, 10), 157(5), *157,* 160 (6), 161(6), 162(6), 163 (10), 164(10, 11), 165 (11), 166(21), 167(21), 168(21), 169(21), 170(21), 175(38), *181, 182,* 184(2, 4, 5, 6, 7), 185(8), *189,* 205, 206(12), 210, *217*
Muszbek, L., 88(83), 98, *106,* 160(5), *181*

N

Nakamura, M., 87(74), 93(111), *105, 107*
Nakamura, R. M., 55(120), *70*
Nakase, Y., 173(35), 174(35), *182*
Nakayama, H., 24, *66*
Niwa, M., 45, 50(87), 54, 60, *68,* 87(74, 75, 76, 77), *105, 106,* 157(23), *158*
Nixon, C., 9(35), *12,* 88(82), 98(82), *106,* 160(4), *181*
North, E. A., 39(68, 69), *68*

O

Ochiai, T., 135(25), 137(27), 139(27), *140, 141*
Okonogi, T., 20, 21(12), 23, *65*
Okumura, K., 135(25), 137(27), 139(27), *140, 141*
Olitsky, P. K., 143, *150*
Olivera Lima, A., 91(100), *107*
Onoue, K., 20, 36, 38, 51(11), *65, 67*

AUTHOR INDEX

Orlando, R. A., 170(27), *182*
Ortez, R. A., 100, *107*
Ospeck, A. G., 21(16), 25(16), *66, 72, 103,* 110, *121*

P

Parfentjev, I. A., 9(33), *12, 48* (100), 49(100, 110), *69, 72,* 73(4, 5, 6, 16, 24, 27, 28, 34), 74(53), 76 (50, 58), 77(50), 88(86), *103, 104, 105, 105,* 110, *121,* 160, 165, 166(20), *181, 182*
Parker, C. W., 45, 54(91), *68,* 100(121), *107,* 153(12), 157(12), *157,* 184(3), *183*
Parton, R., 20, *65*
Patrenco, H., 110(5), *121*
Peixoto, J. M., 124(7), 130(7), *140*
Pekárek, J., 93(110), *107*
Pericin, C., 113(16), *121*
Perkins, F. T., 3(13), 4(13), *11,* 166(17), *182*
Perla, D., 93(106), *107*
Phillips, J. M., 139(31), *141,* 186(11), *190*
Phillips, R. E., 198, *216*
Pieroni, R. E., 31(34), 45, 50 (112), 54, *66, 68, 69,* 73(35, 42), 76(42, 51), *104, 105,* 145(5), 146(5), 147(5), *150,* 210(14), *217*
Pillemer, L., 42(75), 45, *68*
Pincus, G., 93(107), *107*
Pittman, M., 2(10), 3(10), 8 (23), 9(30), *11, 12,* 17 (3), 24(22), 34, 38(53, 55), 39(3), 44(3), 46(3), 48, 49(101, 108), *65, 65, 67, 69,* 73(14, 18), 74 (47), 75(62, 64), 76(47), 82(69), 88(47), 91(18), *103, 104, 105,* 115, *122,* 160(2), *181,* 186(14), *190,* 195, 207(13), *216, 217*

Plager, L., 128(18), 129(19), *140*
Pozsgi, N., 151(3), *157*
Preston, N. W., 35, 39(63), 44, *67, 68*
Prouvost-Danon, A., 130(21), 131, *140*
Ptak, W., 139(28), *141*
Pusztai, Z., 82(68), *105*

Q

Quevillon, M., 73(45), *104*

R

Rapp, H. J., 171(31), *182*
Raška, K., 5(15), *11*
Rasmussen, A. F., 89(92), 94 (92), *106*
Reed, C. E., 139, *141,* 186(12), *190*
Regan, J. C., 7, *11*
Reilly, H. C., 171(29), *182*
Reilly, J., 21(14), *65*
Řežabek, K., 93(110), *107*
Ribi, E., 14(2), 15(2), 20(2), 22(2), 26, 27(32), 28 (32), 29(30), 50(2), *65, 66*
Riester, S. K., 154(17), *157*
Rivalier, E., 21(14), *65*
Rizzo, N. D., 8, *11*
Roberts, M. E., 21(16), 25(16), *66, 72, 103,* 110, *121*
Rochas, S., 130(21), 131(21), *140*
Rose, N. R., 148(14), *150*
Rosett, W., 156(19), *157*
Ross, L. L., 73(22), *104*
Ross, R., 18(6), 19(6), 27(6), 30(6), 33(44), *65, 67*
Rowen, M., 73(37), *104*
Rowley, D. A., 176, 177, *182*
Rudbach, J. A., 26(30), 27(30), 29(30), *66*

S

Sagin, J. F., 32(39), 35(39), 36(39), 38(39), *66*
Samter, M., 73(37), *104*
Sato, Y., 43(76), 45, 49(76), 50(89), 51(76), 55, 60, 64, *68, 69,* 153(14), 157(14), *157,* 188(23), *190*
Sauer, L., 2, 7, *11,* 44, *68*
Schachter, M., 8, *11*
Schaedler, R. W., 73(25), *104,* 166, *182*
Schayer, R. W., 88(78, 79, 80), *106*
Scherp, H. W., 151(2), *157*
Schleyer, W. L., 88(86), *106,* 160, *181*
Schmidt, H., 128(16), *140*
Schobert, L., 9(35), *12,* 88(82), 98(82), *106,* 160(4), *181*
Schuchardt, L. F., 32(39), 35(39), 36(39), 38(39), 49(103), *66, 69,* 73(13, 15, 39), 74(52), 75(65), 76(52), 79(65), 93(39, 102, 105, 108), *103, 104, 105, 107,* 115(21, 22), 116(22), 118(21, 22, 27), *122,* 164(11), 165(11), *181,* 184(4), *189,* 206(12), *217*
Schullenberger, C. C., 170(26), 171(28), *182*
Schweinberg, H., 36(50), *67*
Shirato, E., 170(26), *182*
Sinkovics, J. G., 170(26), 171(28), *182*
Sleeswyk, 1(5), *11,* 32, *66*
Smith, R. F., 51(113), *69,* 111(12), 113(12), *121,* 175(38), *182*
Smolens, J., 35, *67*
Solotorovsky, M., 110(3), 115(18), *121*
Sowinski, R., 144(2), 145(6), *150,* 175, 176(37), *182*
Spink, W. W., 49(104), *69,* 73(29, 40), *104*
Stagner, J. I., 173(33), *182*
Stainer, D. W., 46(95), *69*

Stanbridge, T. N., 35, 39(62), *67, 68*
Standfast, A. F. B., 25(24), *66,* 186(13, 15), *190*
Stevens, W. K., 187(17), *190*
Stiffel, C., 171(30), *182*
Strean, L. P., 24(23), *66*
Ström, J., 8(21), 9(21), *11*
Strong, M. G., 160(2), *181*
Stronk, M. G., 74(47), 76(47), 88(47), *104*
Suzuki, K., 45(89), 49(89, 111), 50(89), 51(111), 55(89, 111), *68,* 188(23), *190*
Svensson, H., 59, *70,* 216, *217*
Szentivanyi, A., 73(44), 76(49), 88(81, 84, 87), 89(84), 94(49, 84, 115), 96(49), 97(115), 100(120), *104, 105, 106, 107,* 150(19), *150,* 160(8), *181,* 184(1), 187(22), 189(22), *189, 190*

T

Tabachnick, I. A., 9(35), *12,* 88(82, 85), 98(82), *106,* 160(4), *181*
Tada, T., 135(25), 137(27), 139(27), *140, 141*
Tajima, M., 166(18), *182,* 186, *190*
Talmage, D. W., 76(49), 88(87), 94(49, 115), 96(49), 97(115), *105, 106, 107,* 160(8), *181,* 184(1), *189*
Tan, E. M., 45(90), 51(90, 114), 55(90, 120), 60(90), *68, 69, 70,* 98(119), *107,* 157(22), *158*
Tanaka, T., 171(31), *182*
Taub, R. N., 139(35), *141,* 156(19), *157*
Taylor, R. B., 139(33), *141*
Teissier, P., 21(14), *65*
Thiele, E. H., 42(73), *68*
Thimann, K. V., 93(107), *107*
Thow, D. W., 186(13), *190*

AUTHOR INDEX

Timmis, C., 9(28), *11*
Tinker, M. R., 145(8), *150,* 151(2), *157*
Tokuda, S., 73(23), 91, 93(23), *104,* 116(23), 120(33), *122*
Tolstoouhov, A., 7, *11*
Torrigiani, G., 139(32), *141*
Townley, R. G., 73(44), *104*
Trapani, I. L., 73(44), *104*
Treadwell, P. E., 89(92), 94(92), *106*
Twarog, F. J., 148(14), *150*

U

Unanue, E. R., 129, *140*
Ungar, J., 187(17), *190*

V

van Hemert, P., 45(94), *69*
Vaughan, J. H., 45(90), 51(90, 114), 55(90), 60(90), *68, 69,* 98(119), *107,* 157(22), *158*
Verwey, W. F., 32(39), 35(39), 36(39), 38(39), 42(84), *66, 68,* 73(13, 15), 75(65), 79(65), 93(108), *103, 105, 107,* 115(21, 22), 116(22), 118(21, 22), *122,* 206(12), *217*
Virion, M. E., 72(4, 5, 6), 73(4, 5, 6), *103,* 110(9), *121*

W

Warburton, M. F., 39(68), *68*
Wardlaw, A. C., 20, 51(115, 116), *65, 69*
Warner, N. L., 202, *217*
Watanabe, Y., 42(72), *68*

Weiner, S. L., 145(8), *150*
Weiser, R. S., 73(23), 91, 93(23), *104,* 115(17), 116(23), 120(33), *121, 122*
Wenk, E. J., 50(112), *69,* 145(4, 5, 7, 9), 146(5, 9), 147(5, 10), *150,* 210(14), *217*
Westphal, O., 26, *66,* 126, *140*
Wheeler, A. H., 110(5), *121*
Wheeler, M. W., 193, *216*
Willard, C. Y., 43(81), *68*
Wilson, G. S., 9(27), *11,* 110(1), *121*
Winsten, S., 110(3), 115(18), *121*
Wissler, R. W., 170(27), *182*
Wistar, R., 89(92), 94(92), *106*
Wood, M. L., 23, *66*
Woods, E. F., 87(71), *105*
World Health Organization, 45, *69*
Wortis, H. H., 128(15), 129(15), 139(15, 33), *140, 141*

Y

Yamadeya, Y., 87(75, 76, 77), *106*
Yamamoto, A., 187(18), *190*
Yamamura, Y., 20(11), 36(51), 38(51), 51(11), *65, 67*
Yoo, T. J., 173, *182*
Yoshida, F., 173(35), 174(35), *182*
Yoshikawa, T., 154(16), 155(16), 156(16), *157,* 187(20), *190*

Z

Zakharova, M. S., 3(14), 5(14), *11*
Zbar, B., 171(31), *182*

Subject Index

A

Adjuvant action:
  of *B. pertussis,* 123, 124, 138-139
  of pertussigen, 185, 186
  mechanism, 139, 185
Adjuvant arthritis, pertussigen effect on, 39
Adrenalectomy in mice, 118
Adrenal gland:
  B. pertussis effect on, 93
  and cold stress, 164
  and hypoproteinemia, 163
  pertussigen effect on, 94, 118
  and shock, 94
Adrenal hormones:
  and histamine shock, 94
  pertussigen effect on, 93
Adrenergic agonists and permeability, 101
Adrenergic blockade:
  and histamine shock, 96
  in shock, 96
  and vascular permeability, 101
Adrenergic blocking agents and histamine shock, 94
Adrenergic receptors:
  and histamine shock, 98
  metabolic effects, 94, 95, 98
  and pertussigen, 97
  and shock, 98
Agglutination of erythrocytes, 200
Agglutination test, 207, 208
Agglutinins in mice, 208

Agglutinogen absorption test, 208, 209
Agglutinogens, 32-39
  antigenic complexity, 32
  antigenicity, 38
  biological activities, 38
  chemical nature, 37
  culture phases, 32
  factors, 33, 35
  general characteristics, 32
  and mouse protection, 38, 39
  mutational changes, 34, 35, 39
  and pertussis vaccine, 39
  properties, 33, 34
  purification, 35-37
  role in immunity, 38
Anaphylaxis:
  active, 110, 120, 206
  and antibody titers, 112
  and body temperature, 120, 121
  in *B. pertussis*-treated mice, 110, 206
  effect of drugs on, 93, 113, 120
  effect of pertussigen, 112, 113, 186
  and Freund's adjuvant, 112
  general remarks, 109
  hematocrit in, 90
  and histamine shock, 93, 116, 120
  mechanism, 118-120
  mediators of, 91, 93
  in mice, 90, 110
  passive, 114, 115, 206
  symptoms of, 119

[Anaphylaxis]
  and vascular bed, 90, 119
Animal species, response to pertussigen, 73
Antibody determination, bis-diazo-benzidine technique, 200-202
Antibody forming cells, enumeration of, 202-205
Antibody response:
  effect of B. pertussis on, 128
  effect of endotoxin on, 126, 128
  effect of lipid A on, 126
  IgE, 209
Antisera, preparation of, 206-207
Autoimmune disease:
  effect of B. pertussis on, 143
  effect of pertussigen on, 185
5'-Adenosine monophosphate and histamine shock, 98

B

Beta receptors and pertussigen action, 184
Bordet-Gengou agar, 193
Bordet-Gengou bacillus (see Bordetella pertussis)
Bordetella:
  bronchiseptica, 2
  bronchiseptica, lipopolysaccharide (endotoxin), 26, 27
  parapertussis, 2
  parapertussis, lipopolysaccharide (endotoxin), 27
  pertussis, 2
Bordetella pertussis:
  adjuvant action, 120, 138-139
  antigens of, 18, 20
  ascites stimulation by, 177, 180
  culture medium for, 193
  culture phases, 18
  diluent for challenge suspensions, 194
  effect on anaphylaxis, 110

[Bordetella pertussis]
  effect on antibody response, 123, 124, 128
  effect on autoimmune disease, 143
  effect on bacterial infections, 165, 166
  effect on cold stress, 164, 165
  effect on delayed hypersensitivity, 176-181
  effect on experimental allergic encephalomyelitis, 145
  effect on fungal infections, 169
  effect on immunoglobulin response, 128, 129
  effect on infection in animals, 9
  effect on insulin levels, 160
  effect on macrophage response to injury, 175, 176
  effect on passive anaphylaxis, 115, 116
  effect on rabies, 166-168
  effect on shock, 72
  effect on tumors, 170-173
  effect on viral infections, 166-168
  effect on viral vaccines, 168
  and enhancement of tumors, 170, 171
  future work on, 188, 189
  general description, 14, 18
  growth requirements, 14, 17
  induction of hypoglycemia, 160
  induction of hypoproteinemia, 162, 187
  induction of interferon, 173, 174
  induction of leukocytosis, 152, 153
  induction of lung edema, 180, 181
  induction of lymphocytosis, 151-153
  lipopolysaccharides (endotoxin), 26, 126, 166
  liquid medium for, 193
  mouse protection test for, 195, 196

SUBJECT INDEX

[*Bordetella pertussis*]
  phases of, 2, 18
  preparation of cell suspensions, 194
  rough form, properties of, 18
  smooth form, properties of, 18
  suppression of cyclic adenosine monophosphate (c-AMP), 100
  and tumor suppression, 171-173

C

Candida albicans infection, effect of *B. pertussis* on, 169
Catecholamines effect on permeability, 101
Circulatory shock, characteristics of, 72
Cryptococcus infection, effect of *B. pertussis* on, 169
Cyclic adenosine monophosphate (c-AMP):
  B. pertussis effect on, 100
  histamine shock and, 98
  pertussigen effect on, 100

D

Delayed hypersensitivity:
  pertussigen effect on, 137
  to pertussis vaccine, 176, 177
Disc electrophoresis in acrylamide gel, 210-214

E

Endotoxin (*see also* Heat stable toxin):
  effect on antibody response, 125, 126, 128
  effect on hypoproteinemia, 163
  effect on immunoglobulin response, 128
  effect on spleen size, 154
  induction of interferon by, 173, 174
  and mouse protection, 166, 137

[Endotoxin]
  nonspecific protection by, 166
  stimulation of IgE, 131
Epinephrine and pertussigen action, 184
Experimental allergic encephalomyelitis (EAE):
  effect of *B. pertussis* on, 210
  effect of *B. pertussis* vaccine on, 145
  effect of Freund's adjuvant on, 145
  hyperacute type, 145-147
    effect of pertussigen on, 147, 148
    effect of pertussis vaccine on, 145
  in Lewis rats, 145
  mechanism of, 148, 185
  passive transfer of, 147
  permeability in, 149, 150
  in mice, 145

F

Freund's adjuvant, 124
  effect on anaphylaxis, 112
  effect on EAE, 145
  effect on IgE response, 131

G

Gel diffusion test, 211
Gram-negative bacteria, infection by, 165, 166

H

Haemophilus pertussis (see *Bordetella pertussis*)
Heat labile toxin, 20-25
  antibodies to, 21
  antigenicity, 21, 24
  effect on EAE, 22
  effect on lymph nodes, 24
  effect on mouse protection, 24
  effect on spleen, 23, 24
  general characteristics, 20

[Heat labile toxin]
  interferon induction, 173, 174
  nature, 20
  purification, 20
  role in immunity, 24, 25
  stability, 21, 22
  toxicity, 22-24
  toxoid formation, 21
Heat stable toxin (see also Endotoxin), 26-32
  agglutinogens, 30
  antigenicity, 27
  general characteristics, 26
  biological activities, 27, 30, 31
  chemical nature, 27
  and histamine hypersensitivity, 31
  purification, 26, 27
  role in immunity, 30
Hemagglutinin, 39-43
  antigenicity, 41
  biological activities, 43
  effect on erythrocytes, 42
  general characteristics, 39, 40
  and mouse protection, 42, 43
  in prophylaxis of whooping cough, 42
  purification, 41
  stability, 42
  toxicity, 41
Histamine sensitivity, effect of hypoglycemia on, 160, 161
Histamine sensitivity test, 196, 197
Histamine sensitization with live B. pertussis, 82
Histamine sensitizing factor (see also Pertussigen and lymphocyte promoting factor):
  effect of enzymes on, 51
  interferon induction, 173, 174
  molecular weight, 51
  purification, 53-65
    Lehrer's method, 55, 56
    Munoz' method, 56-64
    Niwa's method, 54
    Pieroni's method, 54
Histamine shock, role of insulin on, 98

Hydroxylapatite column chromatography, 215
Hypersensitivity to shock:
  duration after pertussigen, 83
  effect of dose of pertussigen on, 83
Hypoglycemia, 160-162
  and histamine sensitivity, 160
  effect of pertussigen on, 89
  effect of pertussis vaccine, 160
  onset and duration after pertussigen, 160
Hypoproteinemia, 162-164

I

Immunoelectrophoresis (IEP) technique, 211, 212
Immunoglobulin E:
  booster effect of pertussigen on, 133
  characteristics, 130
  effect of B. pertussis on, 130
  effect of pertussigen on, 113, 133, 135
  effect of pertussis vaccine on, 124
  passive anaphylaxis produced by, 116
  PCA produced by, 116
  in peritoneal fluid, 132
  pertussigen stimulation of, 130
  with specificity to B. pertussis, 133
Immunoglobulin G_1:
  passive anaphylaxis produced by, 116
  PCA produced by, 116
Insulin effect on histamine shock, 98

L

Leukocyte count, 197, 198
Lipid A, effect on antibody response, 126

SUBJECT INDEX

Lipopolysaccharide (see also Heat stable toxin, Endotoxin), 26, 27, 126, 166
Lung edema, induction by B. pertussis, 180, 181
Lymphocyte promoting factor (LPF) purification, 54-64
  Iwasa's method, 62, 63
  Morse's method, 54, 55
  Sato's method, 55
Lymphocytosis, 151-157
  depletion of lymphocytes from lymphoid tissue, 155, 156
  effect of agglutinins to B. pertussis on, 156
  effect of heating pertussis vaccine on induction of, 153
  effect of reticuloendothelial blockade, 156
  effect of route of administration of pertussis vaccine, 153
  effect of various treatments on, 156
  mechanism, 154-157
  migration of lymphocytes, 156
Lymphoid cells:
  in EAE, 147
  in hyperacute EAE, 147

M

Macrophages, effect of B. pertussis on, 129, 130
Mast cell sensitizing antibody (see also Immunoglobulin E), 130
Mouse, resistance to histamine, 78
Mouse protection
  effect of agglutinogens on, 38
  effect of endotoxin on, 186, 187
  effect of pertussigen on, 186, 187
  and permeability of blood-brain barrier, 187
Mouse protective antigen and pertussigen, 47

Mouse protection test, 9, 195, 196
Mouse strains, response to pertussigen, 74

P

Passive anaphylaxis, 114-116, 120
  antibody class in, 115, 116
  IgE in, 116
  IgG_1 in, 116
Passive cutaneous anaphylaxis, 132
  effect of antihistamines on, 120
  effect of lysergic acid diethylamide (LSD) on, 120
  effect of pertussigen, 118
  induced by IgE, 116
  induced by IgG_1, 116
  in mice, 205, 206
Permeability (see Vascular permeability)
Pertussigen, 18, 47-65
  activities, 57, 59, 62, 63, 65
  adjuvant action of, 185, 186
  adrenergic blockade produced by, 94, 97, 100, 120
  antigenicity, 64
  binding to cells, 156
  and catecholamines, 94
  chemical nature, 51, 57, 59, 64
  dehydration and action of, 78, 79, 81
  diet and action of, 75
  effect of age of mouse, 74
  effect of enzymes on, 51
  effect of mouse strain on action of, 74, 118
  effect of route of administration, 81, 83
  effect of sex of mice on activity, 75
  effect of stressing mice on activity, 75
  effect of toxic substances on activity, 75

[Pertussigen]
  effect on active anaphylaxis, 116
  effect on adrenal gland, 118
  effect on adjuvant arthritis, 138
  effect on albumin levels in plasma, 100
  effect on anaphylaxis, 110, 112, 113
  effect on antibody response, 125
  effect on c-AMP, 100
  effect on cold stress, 164
  effect on delayed hypersensitivity, 177
  effect on different animals, 73
  effect on hyperacute EAE, 147, 148, 185
  effect on hypersensitivity, 48, 50
  effect on insulin levels, 88, 89, 98
  effect on macrophage response to injury, 175, 176
  effect on Schultz-Dale, 118
  effect on serotonin sensitivity, 83
  effect on shock, 72, 120, 184
  effect on spleen, 175
  effect on vasoactive amines, 184
  and endotoxin, 83
  and epinephrine function, 184
  general properties of, 47, 49-51, 53
  and hemagglutinin, 64
  and histamine dose, 75
  and histamine levels, 87, 88
  and histamine sensitivity, 83
  and histidine decarboxylase levels, 88
  historical notes, 48
  housing conditions and action of, 75
  immediate hypersensitivity to, 177
  induction of edema by, 177
  induction of hypoglycemia, 89, 160, 187

[Pertussigen]
  induction of hypoproteinemia, 163
  induction of IgE, 113, 130, 132, 133, 135
  induction of lymphocytosis by, 151, 155, 157, 187
  induction of vascular permeability, 184
  mechanism of action in anaphylaxis, 186
  mode of action, 87, 93, 96, 98, 100-103, 148, 149, 183-187
  molecular weight, 51
  and mouse protection, 45, 186, 187
  passive anaphylaxis, 116
  passive cutaneous anaphylaxis (PCA), 118
  permeability, vascular, 100, 101
  physical form, 81
  purification, 53-64
  purification, Munoz' method, 54, 56-64
  relationship to other antigens, 49
  and reticuloendothelial system, 89
  suppression of antibody response, 135-136
  suppression of delayed hypersensitivity, 136
  toxicity, 51
Pertussis (*see* Whooping cough)
Pertussis vaccine:
  adjuvant action, 123, 124
  changes induced by, 9, 10
  effect of aluminum gels on, 46
  effect on animals, 9
  effect on antibody response, 127
  effect in man, 9
  effect on organs, 154, 155
  effect on passive transfer of hyperacute EAE, 147
  effect of route of administration, 153
  induction of adrenalitis, 148

SUBJECT INDEX

[Pertussis vaccine]
  induction of aspermatogenesis, 148
  induction of hyperacute EAE, 146
  induction of lymphocytosis, 151-154
  induction of thyroiditis, 148
  preparation, 45-47
  preservatives, 46
Plasma glucose determination, 198
Protection test:
  in marmoset, 39
  in mice, 195, 196
Protective antigen:
  and agglutinogens, 44
  general remarks, 43, 44
  histamine sensitizing factor, 45
  induction of lymphocytosis, 45
  interferon induction, 173, 174
  and mouse protection test, 44
  nature of, 44
  purification, 44, 45

S

Schultz-Dale reaction:
  effect of B. pertussis on, 113
  method to demonstrate, 210, 211
Serotonin-histamine sensitivity test, 197
Serum protein determination, 195
Shock:
  anaphylactic, 72, 73
  effect of epinephrine on, 90
  induction by histamine, 73
  in the mouse, 89, 90
  effect of pertussigen on, 184
  induced by serotonin, 73
  effect of various factors on, 73
Starch block electrophoresis, 215, 216
Steroids, effect on histamine shock, 93

T

Thymusless mice, effect of B. pertussis on, 129
Tumors, effect of B. pertussis on, 170-173

V

Vaccination, complications after, 8
Vaccine for whooping cough, 2, 45-47
Vascular permeability:
  and albumin levels in blood, 100
  in anaphylaxis, 119
  and catecholamine, 101
  effect of pertussigen on, 184
  Evans blue dye test for, 198
  in hyperacute EAE, 146
  radiolabeled albumin test for, 199, 200
  tests for, 198-200
Vasoactive amines, effect of pertussigen on, 184

W

Whooping cough:
  blood sugar in, 7
  complications in, 8
  historical notes, 1, 2
  incidence, 2, 3, 5
  leukocytosis in, 151
  lymphocytosis in, 7, 151
  mortality in, 2, 3, 5, 6
  pathology of, 6-8
  physiological changes in, 6
  skin test for, 38
  symptomatology of, 6
  vaccine for, 2, 45-47

Z

Zonal density gradient electrophoresis, 216